Red Sanders: Silviculture and Conservation

T. Pullaiah • S. Balasubramanya
M. Anuradha
Editors

Red Sanders: Silviculture and Conservation

Springer

Editors
T. Pullaiah
Department of Botany
Sri Krishnadevaraya University
Anantapur, Andhra Pradesh, India

M. Anuradha
Department of Biotechnology
Padmashree Institute of Management and Sciences
Bangalore, Karnataka, India

S. Balasubramanya
Department of IT, BT and S&T, Government of Karnataka
Karnataka Innovation & Technology Society
Bangalore, Karnataka, India

ISBN 978-981-13-7629-0 ISBN 978-981-13-7627-6 (eBook)
https://doi.org/10.1007/978-981-13-7627-6

This Springer imprint is published by the registered company Springer Nature Singapore Pte Ltd.
The registered company address is: 152 Beach Road, #21-01/04 Gateway East, Singapore 189721, Singapore

Contents

About the Editors

T. Pullaiah, Ph.D. is a former Professor at the Department of Botany at Sri Krishnadevaraya University, Andhra Pradesh, India, where he has taught for more than 35 years. He has held several positions at the university. He was President of the Indian Botanical Society (2014), President of the Indian Association for Angiosperm Taxonomy (2013), and Fellow of Andhra Pradesh Akademi of Sciences. He was awarded the Panchanan Maheshwari Gold Medal, the Dr. G. Panigrahi Memorial Lecture Award of the Indian Botanical Society, the Prof. Y. D. Tyagi Gold Medal of the Indian Association for Angiosperm Taxonomy, and a Best Teacher Award from the Government of Andhra Pradesh. Under his guidance, 54 students obtained their doctoral degrees. He has authored 53 books, edited 18 books, and published over 330 research papers, including reviews and book chapters. He was also a member of the Species Survival Commission of the International Union for Conservation of Nature (IUCN).

M. Anuradha is a Postgraduate in Botany with university first Rank from Sri Krishnadevaraya University, Andhra Pradesh, India. She continued her M.Phil. as UGC JRF and completed Ph.D. as CSIR SRF, and got doctorate in plant tissue culture of Legume trees. She was a Scientist at Dr. Reddy's Laboratories, Medicinal Plant Research Division, Hyderabad, where she worked in anticancer plants and in vitro production of Taxol and Camptothecins. Later she joined Sami Labs Pvt. Ltd., Bangalore, a reputed herbal and nutraceutical industry, where she worked on various aspects of medicinal and nutraceutical herbs. Along with like-minded technocrats, she started her own company Rishi Herbal Technologies Pvt., Ltd and is a Founder Member and President of Rishi Foundation, Research Institute for Social and Health improvement. She is associated as a Principal in Padmashree Institute of Management and Sciences, playing a key role and bridging gap between industry and academia. She was also Chairperson to Board of Studies to KSOU and Bangalore University and Director to DDU Kaushal Kendra at Padmashree Institute of Management and Sciences. To her credit, she was selected as young scientist by DST, New Delhi, in 2001. Till now she completed projects worth more than 2 crores, from various national and international agencies like Department of Biotechnology, Department of Science and Technology, National Medicinal Plant Board, New Delhi, and Rufford foundation, UK. She is instrumental in starting Biotech finishing school at Padmashree Institute of Management and Sciences, Bengaluru. To her credit, she published more than 30 papers in national and international journals and co-authored book chapters. She also filed six Indian patents and also one US patent, which is granted. She has guided 6 doctoral students and more than 100 student projects. Mentored Department and Science and Technology, Rajat Jayanthi student projects and Vision Group of Science and Technology, spice projects. She travelled extensively to rural and semiurban areas to popularize science and create awareness.

S. Balasubramanya obtained his M.Sc. and Ph.D. degrees in Botany from Bangalore University. He worked in various organizations and companies which include Sami Labs Ltd., Bengaluru, Rishi Herbal Technologies Pvt. Ltd., and Padmashree Institute of Management and Science, Bengaluru. Currently, he is senior consultant for Karnataka Biotechnology and Information Technology Services (KBITS), a State Government organization. He published 13 papers in journals of International repute and contributed a book chapter. He got a US patent on "Enhancing Secondary Metabolites and Nutrients in Plants and Plant Parts" and filed six more patents. He was Principal Investigator/Co-investigator of five major research projects funded by National Medicinal Plants Board, Department of Science and Technology, Government of India, Rufford Foundation, UK, and Charms Chem Pvt. Ltd., India, worth more than a crore of rupees.

Introduction

1

T. Pullaiah and M. Anuradha

Abstract

Pterocarpus santalinus L.f., popularly known as red sanders, belonging to the family Fabaceae, is endemic to the southern parts of Eastern Ghats. IUCN has listed this plant as endangered. The plant has superlative characteristics in its wood and has many medicinal properties. This plant has attracted the attention of both foresters and layman because of its illegal harvesting and law and order problem. In this chapter the plant is introduced to the reader for its importance and restricted distribution.

Keywords

Red sanders · Red sandalwood · *Pterocarpus santalinus* · Endangered tree

1.1 Introduction

The astonishing gifts of forests to the well-being of humankind are from time immemorial. Forest wealth plays a vivacious role in providing green growth opportunities, food production, mitigation of climate change and primary medical needs. The immense onus of forests in supporting and balancing ecosystems, recycling of natural resources and conservation of biodiversity is uncontestable. In 1990 the world had 4.128 million hectares of forest, and this area declined to 3.999 million hectares by the year 2015 (Global forest resources assessment 2015). Half of these forests are in tropical regions. India ranking sixth, among the

T. Pullaiah (✉)
Department of Botany, Sri Krishnadevaraya University, Anantapur, Andhra Pradesh, India

M. Anuradha
Department of Biotechnology, Padmashree Institute of Management and Sciences, Bangalore, Karnataka, India

T. Pullaiah et al. (eds.), *Red Sanders: Silviculture and Conservation*,
https://doi.org/10.1007/978-981-13-7627-6_1

12 megabiodiversity countries of the globe, is a home to 11% of the world's flora, out of which 33% angiosperms are endemic.

Among the most discussed endemic tree of India is *Pterocarpus santalinus* L.f., popularly known as red sanders, belonging to the family Fabaceae. This precious tree is renowned for its characteristic timbers of exquisite colour, beauty, superlative technical qualities and ranks as one among the finest luxury woods in the world. *Pterocarpus* is derived from the Greek words *pteron* (wing) and *karpos* (fruit), referring to the winged pod, while *santalinus* originates from the Latin *sandal* and *inus* (meaning similar to), i.e., a plant with characteristics similar to Indian sandalwood, *Santalum album* L. (Botanical Survey of India 2012). Red sanders is an endangered and endemic taxon, restricted in distribution to certain forest tracts of peninsular India. The natural occurrence of this is estimated approximately to 0.4 million hectares and confined to various geographical locations of Andhra Pradesh mainly in Seshachalam, Lankamala, Veligonda and Palakonda hill ranges and spreads through the districts of Chittoor, Kadapa, Kurnool, Nellore, Prakasam (Hegde et al. 2012; IFGTB 2011; APFD 2011; Soundararajan and Joshi 2012) and Chengalpattu and Vellore districts of Tamil Nadu (Archana et al. 2013). Teixera et al. (2018) mentioned that it is also found in the Chinese provinces of Yunnan, Guangdong and Guangxi and on Hainan Island, where it is referred to as zitan (Kaner et al. 2013). But we are of the opinion that this wood must have been imported to China from southern India and not grown in China.

A report submitted to National Biodiversity Authority in 2016 reiterates its limited distribution, and the factors responsible are attributed to climatic conditions, soil and geographical formations resulting in "ecological niche" (IFGTB 2011).

1.2 Vernacular Names

Arabic (Sandal-e-Almar); Bengali (Rakta Chandan); Gujarati (Ratanjali); Hindi (Lal Chandan, Ragat Chandan, Rukhto Chandan and Undum); Urdu (Sandal); Kannada (Agslue and Honne); Marathi (Tambada Chandana); Oriya (Rakta Chandan); Malayalam (Patrangam, Tilaparni and Rakta Chandanam); Sanskrit (Rakta Chandan); Sinhala (Rath Sandun); Tamil (Chenjandanam, Semmaram and Sivappu Chandanam); Telugu (Erra Chandanam, Rakta Chandanam and Agaru gandhamu)

Pterocarpus santalinus L.f. [Syn.: *Ligonium santalinus* (L.f.) O. Kuntze] is popularly known as red sanders and red sandalwood. The other names of Red Chandan are Rakta Chandan, red sander wood, Santali Lignum Rubrum, red Saunders, Sandalo Rojo, Rubywood, Sandalwood Padauk, Santalum Rubrum, Sappan, *Pterocarpus santalinus* and Bois de Santal Rouge (Reddy 2018).

The wood has historically been valued in China, particularly during the Ming Dynasty and Qing Dynasty periods, and is referred to in Chinese as zitan (紫檀) and spelt tzu–t'an by earlier western authors such Gustav Ecke, who introduced classical Chinese furniture to the west. Between the seventeenth and nineteenth centuries in China, the rarity of this wood led to the reservation of zitan furniture for the Qing dynasty imperial household. Chandan, the Indian word for red sandalwood which is

Fig. 1.1 Red sander logs. (Courtesy Reddy 2018 with permission)

tzu–t'an, is linked by etymology. The Sanskrit Chandan is in turn etymologically derived from dravidian caandu. The word tan in Chinese is a perfect homonym of "tan", meaning cinnabar, vermillion, and the cognition is suggested by the interchange of Chan for oriflamme, the vermilion ensign of the ancients. Chinese traders would have been familiar with Chandan. Tzu–t'an then is the ancient Chinese interpretation for the Indian word Chandan for red sandalwood. In India red sander wood is one main and lucrative market for smugglers, as a high price is paid for this wood in China. Since the exporting of red sandalwood in India is banned, the underground market is growing, and there are several arrests every year of those trying to smuggle this wood to China (Ramabrahmam and Sujatha 2016).

The wood has been historically valued in China who introduced classical Chinese furniture to the west. It has been one of the most prized woods for millennia. The worldwide market standing is growing in all countries in faster pace as gold. The red sanders is popular for its extremely hard heartwood, and the colour of which is like blood with occasional light yellow streaks. The wood is famous for its medicinal properties and commercial uses (Fig. 1.1).

The red sander wood at the centre of the trunk (heartwood) is mainly used for medicinal purposes. Red sander wood has various medical uses such as treating fluid retention, treating digestive tract problems, purification of blood and treatment of coughs. The red sander wood is of high demand in China, Korea, Singapore, Malaysia, Canada, the USA, the UK, Singapore, Malaysia and Gulf countries (Reddy 2018).

Red sanders contain red coloured pigment santalin and are used in traditional system of medicine against various ailments. Santalin, belonging to the group of flavonoids first reported to have blood red colour, is used in wool dyeing, and alcohol extract of wood is also used for colouring tinctures and varnishes and staining wood (Nietzki 1892). The major constituents of red sander dye are santalin-A, santalin-B and deoxysantalin. Santalin has a quinonoid moiety which is soluble in alcohol yielding a blood red solution with alcohol, yellow with ether and violet with ammonia and caustic alkali (Puri and Seshadri 1955). This has many pharmaceutical, cosmeceutical and nutraceutical applications apart from its usage in wood polish, metal varnish, textiles, wool, silk, hair, leather and jute (Wealth of India 1969). In Europe it is classified as a spice extract rather than food colourant, and in the USA, it is approved as a food dye for alcoholic beverages (Mulliken and Crofton 2008).

The unmatched quality of the timber is due to its gorgeous colour, wavy grain nature and hard, pest-resistant wood and is used for making a musical instrument called shamisen and fetches a fancy price as it produces resonance par excellence. The wood is highly valued for its heavy, dark claret-red heartwood which yields 16% of red colouring matter santalin and is used as colouring agent in pharmaceutical preparations, wine and food. In the traditional system of medicine, the decoction prepared from the heartwood is used in inducing vomiting and treating eye diseases, mental aberrations and ulcers. The heartwood of red sanders is known to have antipyretic, anti-inflammatory, anthelmintic, tonic, haemorrhagic, dysenteric, aphrodisiac and diaphoretic activities (Kirtikar and Basu 1935). Ethanolic extract of stem bark was reported to possess hypoglycaemic activity. The wood in combination with other drugs is also prescribed for snake bites and scorpion stings. Phytochemical investigations of aqueous and ethanolic extracts of stem bark revealed the presence of alkaloids, phenols, saponins, glycosides, flavonoids, triterpenoids, sterols and tannins. The heartwood contains isoflavone glucosides and two antitumour lignans, viz. savinin and calocedrin.

The demand for red sandalwood because of its multi-utilitarian characters is continuously increasing, leading to declination of natural strands, pushing to a fragile situation. Ecologically the species is highly significant as this is an endemic, endangered and red-listed tree, however, having high adaptability to extreme environmental conditions. Exports of red sanders date back to the seventeenth century and are used for textile dying. The exemplarily hard, scarlet-red coloured wood with purple tinge is in enormous demand in China and Japan known to be used in phytomedicine, handicrafts, musical instrument, furniture and idols (Archana et al. 2013). The trade of red sanders is heavily restricted; however illegal felling and smuggling are rampant. The natural strands are an attraction to poachers, and the species is classified as endangered in IUCN Red List of Threatened Plants (Walter and Gillet 1998; Barstow 2018.). IUCN places red sanders under "EN B1 + 2de" category meaning "habitat loss/degradation or extraction or wood or selective logging or clear cutting and the populations are continuing to decline" (IUCN 2007) because of anthropogenic pressures. Latest technological interventions were implemented to map and understand population of red sanders by Giriraj et al. (2008). Its vulnerability and its frequent appearance in flashing news of criminal activities raised alarm among the environmentalists and scientists. Archana et al. (2013) used hyperspectral remote sensing technology and claim that this study can be a ready reference for the forest authorities to monitor cases of mass deforestation.

The elite properties of red sandalwood are attributed to its wavy grain and deep red heartwood which is distinguishable only after 25–30 years of growth. Commercial cultivation is also hampered because of lack of awareness, long juvenile period, lack of cultivation practices, difficult to root nature, mixed population of smooth grain and wavy grain saplings raised from heterozygous seed lot.

Lack of comprehensive information and a myth about the adaptability of the tree to various geographical locations are the major challenges for sustainable utilization of this precious tree. There are publications in various aspects of red sanders (see Red Sanders Information System – wst.icfre.gov.in/database/Redsanders/bibliography/bib.htm); however the information is staggered, and hence a comprehensive monograph on this tree covering taxonomy, morphology, distribution, silvicultural aspects, phytochemistry, pharmacology, propagation, cultivation practices, developmental biology, biotechnology, silviculture, market information, molecular studies, the list of institutions involved in the research and conservation of red sanders and grey areas of research is necessary.

References

APFD (2011) Andhra Pradesh forestry inventory report. Andhra Pradesh Forest Department. Government of Andhra Pradesh, Hyderabad

Archana J, Tanushri S, Swetha M, Ravibabu MV (2013) Identification and mapping of red sandalwood using hyperspectral imagery. Int J Appl Eng Res 8(19):2471–2474

Barstow M (2018) *Pterocarpus santalinus*. The IUCN Red List of Threatened Species 2018: e. T32104A67803072. https://doi.org/10.2305/IUCN.UK.2018-1.RLTS.T32104A67803072.en. Downloaded on 15 October 2018

Botanical Survey of India (2012) Identification protocol for red sanders. In: Pharmacognosy of negative listed plants. Ministry of Environment and Forests, India, pp 171–181. http://wccb.gov.in/WriteReadData/userfiles/file/PROTOCOL%20FOR%20Red%20sanders%20identification.pdf

Giriraj A, Shilpa G, Sudhakar Reddy C, Sudhakar S, Beierkuhnlein C, Murthy MSR (2008) Mapping the geographical distribution of *Pterocarpus santalinus* L.f. (Fabaceae) – an endemic and threatened plant species using Ecological Niche Model. In: Proceeding of the national seminar on conservation of Eastern Ghats. EPTRI, Hyderabad. At Chennai, Volume: 1. www.researchgate.net/figure/Locations-of-Pterocarpus-santalinus-overlaid-on-topography-map-for-the-Eastern-Ghats-of_fig1_262564054

Global Forest Resources Assessment (2015) Food and Agriculture organisation of the United Nations. Available at http://www.fao.org/3/a-i4808e.pdf

Hegde M, Singh BG, Krishnakumar N (2012) Non-detriment findings (NDFs) study for *Pterocarpus santalinus* L.f. (Red Sanders) in India. Available at https://cites.org/sites/default/files/ndf_material/NDF%20Report_Pterocarpus%20santalinus_India.pdf

IFGTB (2011) Collection of quantitative field data through rapid assessment of population, growing stocks and natural regeneration status of *Pterocarpus santalinus* L.f. for CITES non-detriment findings study. A final technical report of completed project submitted to MoEF Government of India. 116 p.

IUCN (2007) *Pterocarpus santalinus*. In: 2007 IUCN Red List of Threatened Species

Kaner J, Jiufang L, Yongji X, Ioras¸ F (2013) A re-evaluation of woods used in Chinese historic furniture (part two). Bull Transilv Univ Brasov Ser II For Wood Ind Agric Food Eng 6:31–40

Kirtikar KR, Basu BD (1935) Indian medicinal plants. International Book Distributors, New Delhi, pp 826–827

Mulliken T, Crofton P (2008) Review of the status, harvest, trade and management of seven Asian CITES-listed medicinal and aromatic plant species. BFN, Skripten 227

Nietzki R (1892) Chemistry of the organic Dye stuffs. Translated with additions by A. Collin and W. Richardson. Santalin. http://scans.library.utoronto.ca/pdf/5/21/chemistryoforgan00nietuoft/hemistryoforgan00nietuoft.pdf. 263 p

Puri B, Seshadri TR (1955) Survey of anthoxanthins. Part IX. Isolation and constitution of palasitrin. J Sci Ind Res 14B:1589–1592

Ramabrahmam V, Sujatha (2016) Red Sanders in Rayalaseema region of Andhra Pradesh: importance to commercial & medicinal value. IOSR J Pharm Biol Sci 11(1):57–60

Red Sanders Information System (*Pterocarpus santalinus* L.) Bibliography wst.icfre.gov.in/database/Redsanders/bibliography/bib.htm

Reddy OSK (2018) Red Sanders – *Pterocarpus santalinus* – Rakta chandan – The Red Gold – Green wealth – an opportunity for Indian Farming community – let's explore. https://www.linkedin.com/pulse/red-sanders-gold-green-wealth-future-generations-indian-osk-reddy/

Selvam ABD (2012) Pharmacognosy of negative listed plants. (Identification Protocol for Red sanders). Ministry of Environment and Forests. Botanical Survey of India. pp 171–181

Soundararajan V, Joshi SC (2012) Endemic possessions of Eastern Ghats: Red Sanders (*Pterocarpus santalinus* Linn.f.). Institute of Wood Science and Technology, Bangalore

Teixeira JA, Kher MM, Soner D, Nataraj M (2018) Red sandalwood (Pterocarpus santalinus L.f.): Biology, importance, propagation and micropropagation. J. For. Res. https://doi.org/1007/s11676-018-0714-6

Walter KS, Gillett HJ (1998) IUCN Red List of Threatened Plants. IUCN – The World Conservation Union: [i]–lxiv, 1–862 p. https://archive.org/details/1997iucnredlisto97walt

Wealth of India (1969) A dictionary of Indian "Raw materials and industrial products"

Taxonomy and Distribution

2

T. Pullaiah

Abstract

Pterocarpus santalinus (red sanders), an endemic and endangered taxon, belongs to the tribe Dalbergieae of the family Fabaceae. There are 46 recognised species in the genus; however red sanders has a special status because of its multi-utilitarian features, endemic nature and supreme technical qualities. This species, though confined to narrow geographical terrains, is introduced and successfully grown in many other areas. In this chapter the detailed taxonomy and distribution are deliberated.

Keywords

Red Sanders · *Pterocarpus santalinus* · Taxonomy · Distribution

2.1 Introduction

The genus *Pterocarpus* belongs to the tribe Dalbergieae of the family Fabaceae, also known as Leguminosae. *Pterocarpus* is a well-recognised genus of trees and woody climbers distributed throughout the world in three tropical regions, i.e. Neotropics, tropical Africa and Indomalaya (Indian subcontinent and Malay Peninsula/Archipelago). The scientific name is Latinized Ancient Greek and means "wing fruit", referring to the unusual shape of the seed pods in this genus. Padauk wood is obtained from several species of *Pterocarpus*. All padauks are of African or Asian origin and are valued for their toughness, stability in use and decorativeness, most having a reddish wood. Most *Pterocarpus* woods contain either water- or alcohol-soluble substances and can be used as dyes.

T. Pullaiah (✉)
Department of Botany, Sri Krishnadevaraya University, Anantapur, Andhra Pradesh, India

T. Pullaiah et al. (eds.), *Red Sanders: Silviculture and Conservation*,
https://doi.org/10.1007/978-981-13-7627-6_2

The genus *Pterocarpus* comprises of 35–40 species (Klitgaard and Lavin 2005) but is estimated to have up to 60 species (Chauhan and Vijendra Rao 2003). *The Plant List* includes 227 scientific plant names of species rank for the genus *Pterocarpus*. Of these 46 are accepted species names (ILDIS and http://www.theplantlist.org/browse/A/Leguminosae/Pterocarpus/ accessed on 2-11-2018).

The 46 species of the genus *Pterocarpus* are listed below:

Pterocarpus acapulcensis Rose
Pterocarpus acuminatus (Graham in Wall.) Kuntze
Pterocarpus albopubescens Hauman
Pterocarpus amazonum (Benth.) Amshoff
Pterocarpus amoenus (Wall.) Kuntze
Pterocarpus angolensis DC.
Pterocarpus antunesii (Taub.) Harms
Pterocarpus brenanii Barbosa & Torre
Pterocarpus claessensii De Wild.
Pterocarpus cuneifolius (Graham) Kuntze
Pterocarpus dalbergioides DC.
Pterocarpus echinatus Pers.
Pterocarpus elegans (Graham) Kuntze
Pterocarpus erinaceus Poir.
Pterocarpus ferrugineus (Roxb.) Kuntze
Pterocarpus forstenianus (Blume ex Miq.) Kuntze
Pterocarpus gilletii De Wild.
Pterocarpus grandifolius (Graham) Kuntze
Pterocarpus heyneanus (Graham) Kuntze
Pterocarpus hockii De Wild.
Pterocarpus homblei De Wild.
Pterocarpus indicus Willd.
Pterocarpus lucens Guill. & Perr.
Pterocarpus macrocarpus Kurz
Pterocarpus marsupium Roxb.
Pterocarpus mildbraedii Harms
Pterocarpus mutondo De Wild.
Pterocarpus officinalis Jacq.
Pterocarpus orbiculatus DC.
Pterocarpus osun Craib
Pterocarpus ovalifolius (Wight & Arn.) Kuntze
Pterocarpus pubipetalus (Miq.) Kuntze
Pterocarpus rohrii Vahl
Pterocarpus rotundifolius (Sond.) Druce
Pterocarpus santalinoides DC.
Pterocarpus santalinus L.f.

Pterocarpus scandens (Roxb.) Kuntze
Pterocarpus soyauxii Taub.
Pterocarpus ternatus Rizzini
Pterocarpus tessmannii Harms
Pterocarpus tinctorius Welw.
Pterocarpus velutinus De Wild.
Pterocarpus villosus (Benth.) Benth.
Pterocarpus violaceus Vogel
Pterocarpus zehntneri Harms
Pterocarpus zenkeri Harms

Most of these species are found in Africa, notably in Nigeria, Cameroon, Sierra Leone and Equatorial Guinea. Four species of *Pterocarpus* – *P. dalbergioides*, *P. indicus*, *P. marsupium* and *P. santalinus* – are found in India. *P. dalbergioides*, an evergreen tree, occurs only in the Andaman (Prasad et al. 2008). *P. indicus*, a native of Malaysia, has been introduced and planted as a garden and avenue tree in the Andaman, West Bengal, Tamil Nadu and Maharashtra. *P. marsupium* is a deciduous tree and occurs commonly in hilly regions throughout the Deccan Peninsula and extends to Gujarat, Madhya Pradesh, Uttar Pradesh, Bihar and Odisha. *P. santalinus* is also a deciduous tree and is restricted to Kadapa (formerly known as Cuddapah), Chittoor, Kurnool and Nellore districts in Andhra Pradesh and Vellore (formerly known as North Arcot) and Chengalpattu districts in Tamil Nadu up to 500 m. It has been reported to be a native of Africa, but its entry into a restricted part of India remains a mystery (Rao and Raju 2002).

Pterocarpus has gained popularity globally for its ethnomedicinal uses and valuable bioactive compounds. The importance of this genus was appreciated by Saslis-Lagoudakis et al. (2011). They reported the application of community phylogenies in bioscreening and shedding light on the processes of shaping cross-cultural ethnomedicinal patterns.

Among the four species mentioned above, the most important species which needs immediate attention and concern is *Pterocarpus santalinus*, an endemic species restricted to Southern parts of Eastern Ghats of India. Being a commercially important tree of Indian forests, it is popularly known as red sanders, red sandalwood, almug and saunderswood. Red sanders is a typical leguminous species of tropical dry deciduous forests of Andhra Pradesh being localized and predominantly confined to Kadapa (Cuddapah) landscape between 13° 30′ and 15° 00′ north latitude and between 78° 45′ and 79° 39′ east longitude (Raju et al. 1999). It is distributed in an area of approximately 5160 km^2 in its geographical spread of scattered forests and in elevated parts of Chittoor, Kadapa (Cuddapah), Kurnool and Nellore districts of Seshachalam hill ranges (150–900 m). It is also found partly distributed in the Vellore (North Arcot) and Chengalpattu districts of Tamil Nadu and in Karnataka (GOI 2014). It is reported that *P. santalinus* occurs in other countries like China, Pakistan, Sri Lanka and Taiwan but must have been an introduced species (Kumar and Sane 2003).

In the area of distribution of red sanders, two species of *Pterocarpus* occur; these are keyed out in the following manner.

Pterocarpus* L. *nom. cons.

1. Flowers in axillary racemes; leaflets usually 3, rarely 4-5, broadly ovate or nearly orbicular; pod concavely curved between the stipe and style; restricted..*P. santalinus*
1. Flowers in terminal panicles; leaflets 5-7, elliptic-oblong; pod convexly curved between the stipe and style; widely distributed..........................*P. marsupium*

Pterocarpus santalinus L. f., Suppl. Pl. 318. 1781; Baker in Hook. f., Fl. Brit. India 2: 239. 1876; Gamble, Fl. Madras 1: 385. 1918.

Tall, deciduous trees, up to 20 m high, bark brownish black, divided into rectangular plates by deep vertical and horizontal cracks, branchlets glabrous. Leaves 3-foliolate, rarely 4-5-foliolate, leaflets ovate-orbicular or oblong, 4–6 × 3.5–5 cm, base obtuse-subcordate, margin entire, apex emarginated; petiole to 4 cm, petiolule to 6 mm. Flowers axillary in simple or sparingly branched racemes; pedicel to 5 mm; calyx tomentose, tube 5 mm, lobes ovate; corolla yellow, standard ovate, 2.1 × 1.5 cm, wings 1.8 × 0.8 cm, keels 1.4 × 0.35 cm; stamens monadelphous (10), staminal sheath 4 × 3 mm, filaments to 5 mm; ovary stipitate, 7 mm, style 4 mm. Pods round with a broad wing, 3–4 cm in diameter, brown; seeds 1 or 2, reddish brown (Figs. 2.2, 2.3, 2.4, 2.5).

2.2 Distribution

Pterocarpus santalinus L.f. (red sanders) is an endemic and endangered species largely confined to the southern portion of the Eastern Ghats, Andhra Pradesh, India.

Red sanders is found distributed in a geobotanically restricted area of 2000 km^2, between 13° 30′–15° 0′ N latitude and 78° 45′–79° 39′ E longitude (Raju et al. 1999). Within this natural distribution, red sanders grows in approximately 500 km^2 of fragmented forest landscape of Andhra Pradesh and Tamil Nadu. The total growing stock of red sanders in AP forests (Fig. 2.1) has been estimated at 118,000 m^3 (Kukrety 2011).

P. santalinus occurs in the forest formation which is called as "5A/C3 Southern Tropical Dry Deciduous Forests" as per Champion and Seth (1968) classification and falls in the eco-terrestrial region IM1301 Deccan Thorn Forests as well as IM0201 Central Deccan Plateau Dry Deciduous Forests. The red sanders is naturally distributed in approximately 3.98 lakh hectares in Seshachalam, Veligonda, Lankamala and Palakonda hill ranges running through five districts, namely, Chittoor, Kadapa, Kurnool, Nellore and Prakasam districts in Andhra Pradesh, India (Pullaiah and Sandhya Rani 1999; Pullaiah and Srirama Murthy 2001, 2018). It also occurs to a small extent in Chengalpattu and Vellore districts in Tamil Nadu. It is endemic to this region. It has been introduced in Sri Lanka, Karnataka and Kerala in the last few decades.

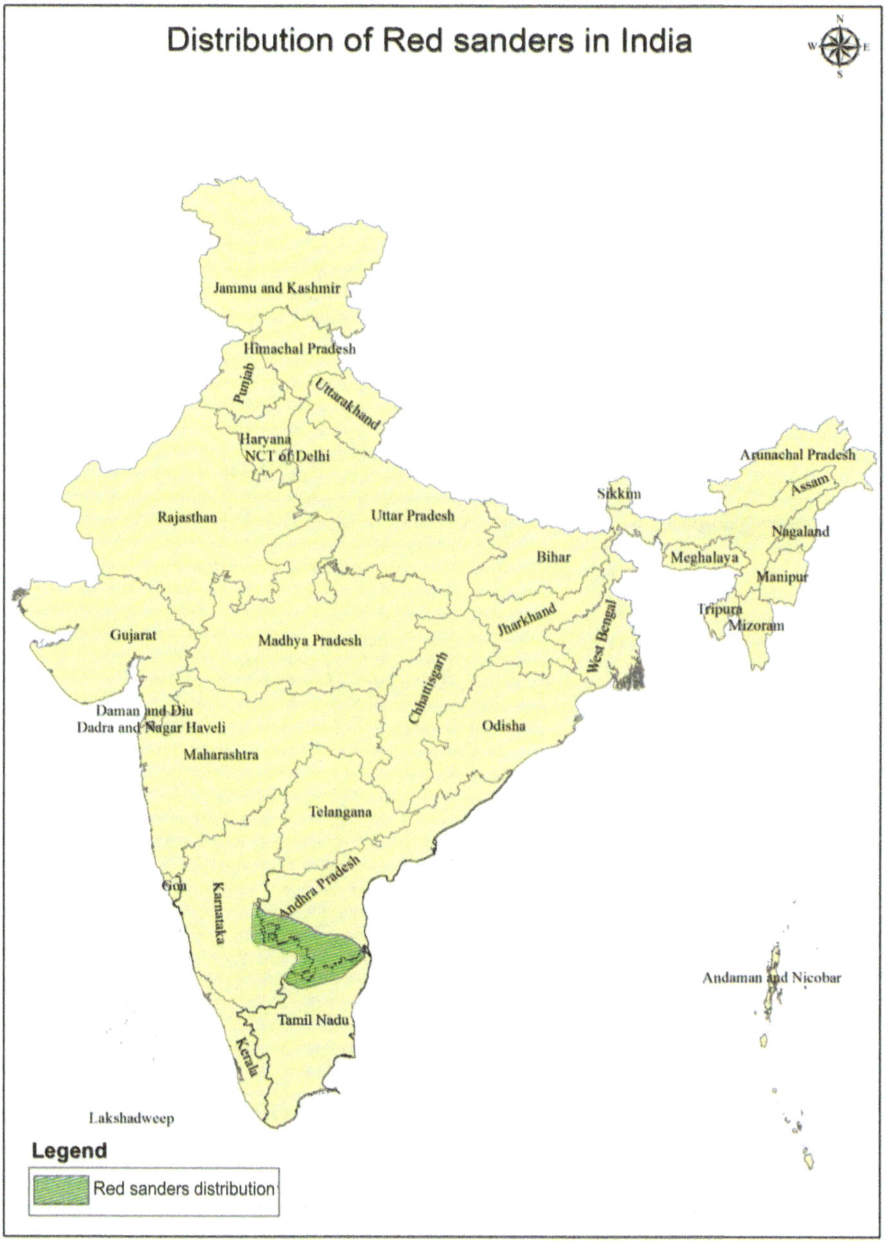

Fig. 2.1 Location map of red sanders in Southern Andhra Pradesh

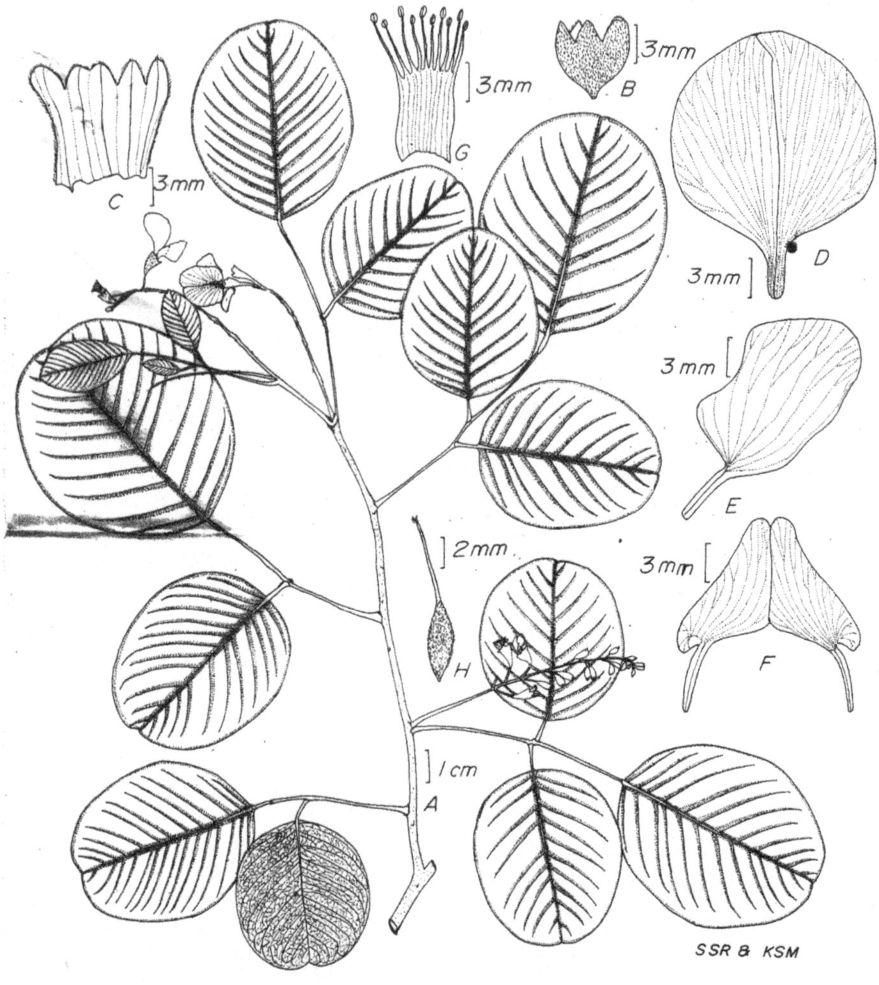

Fig. 2.2 ***Pterocarpus santalinus*** L.f. (**a**)Twig, (**b**) calyx, (**c**) calyx split open, (**d**) standard petal, (**e**) wing petals, (**f**) keel petals, (**g**) staminal column, (**h**) pistil

Specimens Examined Rudravaram – Upper Ahobilam (KNL), *B.Ravi Prasad Rao, T.Shali Shaheb & MO* 29070 (SKU); Rayachoti RF (KDP), *K.Sri Rama Murthy & S. Sandhya Rani* 17647 (SKU); Guvvalacheruvu RF (KDP), *R.V. Reddy* 8170 (SKU); Idupulapaya (KDP), *K.Raja Kullayi Swamy* 40313 (SKU); Tirumala (CTR), *K.Sri Rama Murthy & S.Sandhya Rani* 16020 (SKU); Talakona (CTR), *S.Sandhya Rani* 13732 (SKU); Sidduleswara kona (NLR), *S.Sandhya Rani* 19377 (SKU); Andhra University campus (VSKP), *T.Pullaiah & E.Chennaiah* 7324 (SKU); Rollamadugu (KDP), *J.S.Gamble* 11148 (CAL); Kodur (KDP), *B.H. Anandathirtha Rao* 107361 (DD); Malakonda (NLR), *M.S.Rama Swami* 1255 (CAL).

Chromosome number 2n = 22 (Bhaskar 1981).

Fig. 2.3 Field photographs of *Pterocarpus santalinus*. Images by A. Lalithamba with permission

Fig. 2.4 Leaves, flowers, fruits and wood of *Pterocarpus santalinus*. Images by A. Lalithamba with permission

Fig. 2.5 Bark, wood, wood carvings and wood powder of red sanders. Images by A. Lalithamba with permission

References

Bhaskar V (1981) Chromosome number of *Pterocarpus santalinus* L.f. (Red Sanders). Indian J For 4(4):335

Champion HG, Seth SK (1968) A revised survey of the forest types of India. Government of India, New Delhi

Chauhan L, Vijendra Rao R (2003) Wood anatomy of the legumes of India. Their identification, properties and uses. Bishen Singh Mahendra Pal Singh, Dehra Dun

GOI (Government of India) (2014) Government of India. Ministry of Environment of Forests Publications. Red Sanders. Available at http://moef.nic.in/downloads/public-information/Chap-8-new.pdf. Accessed 24 May 2014

Klitgård B, Lavin M (2005) Dalbergieae *s.l.* In: Lewis G, Schrire B, Mackinder B, Lock M (eds) Legumes of the world. Royal Botanic Gardens, Kew, pp 307–333

Kukrety S (2011) Ecological restoration of Red Sanders (*Pterocarpus santalinus* L.) forests – an adaptive collaborative modelling supported approach. https://www.rufford.org/rsg/projects/sidhanand_kukrety. Accessed 10 Sept 2018

Kumar S, Sane PV (2003) Legumes of South Asia, a check-list. Royal Botanic Gardens, Kew

Prasad RCP, Reddy CS, Raza SH, Dutt CBS (2008) Population structure, age gradations, and regeneration status of *Pterocarpus dalbergioides* Roxb., an endemic species of Andaman Islands, India. Pac J Sci Technol 9:658–664

Pullaiah T, Sandhya Rani S (1999) Trees of Andhra Pradesh. Regency Publications, New Delhi, pp 211–213

Pullaiah T, Srirama Murthy K (2001) Flora of Eastern Ghats, vol 2. Regency Publications, New Delhi, pp 297–298

Pullaiah T, Srirama Murthy K (2018) Flora of Andhra Pradesh, vol 2. Scientific Publishers, Jodhpur, pp 557–559

Raju KK, Raju AN, Sudheer AS (1999) Biogeochemistry of Red Sanders from southeastern part of Andhra Pradesh, south India. Fresenius Environ Bull 8(3–4):204–212

Rao SP, Raju AJS (2002) Pollination ecology of the red sanders *Pterocarpus santalinus* (Fabaceae), an endemic and endangered tree species. Curr Sci 83:1144–1148

Saslis-Lagoudakis CH, Klitgaard BB, Forest F, Francis L, Savolainen V, Williamsons EM, Hawkins JA (2011) The use of phylogeny to interpret cross-cultural patterns in plant use and guide medicinal plant discovery: an example from *Pterocarpus* (Leguminosae). PLoS One 6(7): e22275. https://doi.org/10.1371/journal.pone.0022275. http://www.theplantlist.org/browse/A/Leguminosae/Pterocarpus/. Accessed 2 Nov 2018

3 Wood Uses, Ethnobotany and Pharmacognosy

T. Pullaiah and B. N. Divakara

Abstract

Pterocarpus santalinus, red sanders, an endemic and endangered taxon, is restricted in its distribution to certain forest tracts of southern peninsula of India. Red sanders wood is used in making acoustic instruments, furniture, wood toys, carvings and antique products and in atomic reactors. By virtue of its occurrence in the forests rich in ethnic tribes, the usage of different parts of the tree by these communities is witnessed and documented by researchers. The tree is not only important for its timber but also used for the treatment of many ailments. A comprehensive review on ethnobotany and pharmacognosy is depicted in the current chapter.

Keywords

Red sanders · Pterocarpus santalinus · Wood uses · Shamisen · Zitan · Wood carvings · Pharmacognosy · Ethnobotany

3.1 Wood Uses

Wood is a porous and fibrous structural tissue found in the stems and roots of trees with the matrix of lignin and secondary xylem beneath the bark. The stem wood is bifurcated as sapwood and heartwood. Heartwood is the product of naturally occurring chemical transformation with more resistance to decay, whereas sapwood is the outermost younger wood with living tissues.

T. Pullaiah (✉)
Department of Botany, Sri Krishnadevaraya University, Anantapur, Andhra Pradesh, India

B. N. Divakara
Institute of Wood Science and Technology, ICFRE, Bangalore, Karnataka, India

T. Pullaiah et al. (eds.), *Red Sanders: Silviculture and Conservation*,
https://doi.org/10.1007/978-981-13-7627-6_3

The wood of red sanders has historically been valued in China, particularly during the Ming and Qing periods, referred to in Chinese as *Zitan*, and Chinese traders are familiar with red sanders. This is known for its pretty reddish heartwood which is used in wood works for making antique products.

The wood has historically been valued in China, particularly during the Qing Dynasty periods, and is referred to in Chinese as zitan (紫檀) and spelt tzu-t'an by earlier western authors such as Gustav Ecke, who introduced classical Chinese hardwood furniture to the west. An exquisite chair made of red sandalwood can be seen today in China's Forbidden City in Beijing, inside the Hall of Supreme Harmony, and is once used by the emperors of the Qing Dynasty.

The wood of red sanders is comparatively quite heavy and extremely hard, with the weight of 900–1265 kg per cu m at 12% m.c. and specific gravity with air-dry of 0.87–1.20. The timber has characteristic deep red coloured dye due to the presence of santalin group of compounds, soluble in alcohol. This tree yields an exceptionally strong timber which can be used as house posts. The heartwood of some plus trees is found to have wavy grain, impregnated with deep-red-coloured dye principles. The factors governing the production of wavy grain have not yet been understood as trees with wood of wavy grain occur mixed with other trees in the natural forest. The wood is quite hard, and it is extremely difficult to saw when it is dry. But it is easy to work with hand tools and is highly suitable for carving and turning. It has interlocked grains and required care to be brought to a good finish; however it gets polished very well. Timbers exhibiting wavy grain are highly valuable as much as six times than smooth grained timber. The colour of the red sanders heartwood varies from red to purplish black with interlocked grains, fine textured, strong, in contrast sapwood is white in color (Kumar and Joshi 2014). Lumber pieces with wavy grain margin are considered as superior grade and fetch higher prices (Ramabrahmam and Sujatha, 2016). Artificially coloured saw dusts and wood flakes of another tree, *Adenanthera pavonina* Willd. (Mimosaceae), commonly known as "Raktakambal", are a cheap replacement and sold as an alternative for red sanders (Selvam 2012).

There is no information available based on research with respect to the seasoning qualities of the timber. It has been used extensively since long past as house posts and for carving dolls. Examination of these specimens does not show any signs of cracking or splitting. It is very steady timber and shows very little shrinkage. It may therefore be concluded that the timber seasons well.

It is an extremely durable timber. The heartwood is virtually immune to termite and other insect attacks or to decay. Hence there is no need to consider preservative treatment.

The various uses of wood are listed below.

3.1.1 Historical and Traditional Value

The Ming Dynasty and Qing Dynasty periods introduced classical Chinese furniture to the west. Chinese people believed red sandalwood furniture is lucky at home and is a symbol of status. During the seventeenth and nineteenth centuries in China, zitan furniture was restricted to the Qing dynasty imperial household. The *zitan* and *Huali* with special cultural significance are derived from the woods of *Pterocarpus santalinus*, *Dalbergia odorifera*, *Dalbergia luovelii*, *Dalbergia maritima* and *Dalbergia normandi*. All similar woods are named in trade as rose or violet rosewood which when cut are bright crimson purple changing to dark purple (Ramabrahmam and Sujatha 2016; Mahammadh 2014). In Hinduism, this traditional wood is considered as sacred and used by priests and Brahmins for many rituals and ceremonies, to decorate the idols of the deities and to make religious utensils. Buddhists prefer this wood while cremating. The straight grained wood is used for carving idols, toys and traditional carvings (Ramabrahmam and Sujatha 2016; Wenbin and Xiufang 2013).

3.1.2 Acoustical Instruments

Red sanders is in huge demand in the international market especially in China and Japan for the production of a special musical instrument, *shamisen*, with superior acoustic quality, attributed to the wavy grain of the heartwood (Ramabrahmam and Sujatha 2016; Kumar and Joshi 2014). The musical instrument and articles made from this wood are presented in traditional Japanese wedding as an essential dowry (Ramabrahmam and Sujatha 2016).

3.1.3 Traditional Tirupati Toys and Carvings

Marapachi toys, made of red sanders wood, are essentially displayed during Navaratri/Dussehra as part of Golu or Bommai Kolu festival with a traditional belief of representing them as raja rani dolls, i.e. king and queen. These dolls are also gifted to newly married couple, as a heritage from mother to daughter and for their children. The importance of these gifts is due to the medicinal properties of the wood. The phytocompounds in the wood are ingested by the child while playing considered to be good for health. Red sanders wood is used for delicate carving of mythological stories and figures in the temple and in the palace entrances, sculptures and deities' chariots along with the combination of teak wood if necessary as they are resistant to white ants and fire (Wealth of India 1969).

3.1.4 Atomic Power Plants

Few studies revealed the accumulation of various rare earth elements such as uranium, cadmium, strontium, copper and zinc in the heartwood and leaf of red sanders. The concentration of uranium was higher in *P. santalinus* (Kumar and Joshi 2014).

According to the reports (Jan 19, 2011, 06:47 IST by Poornima Swaminathan, Sandalwood used to cool Chinese nuke reactors?), arrested smugglers revealed/claimed that countries like China, Japan, Myanmar and others in East Asia,

Fig. 3.1 Transverse section of stem of red sanders showing the bark, sapwood and heartwood. (Courtesy Reddy 2018 with permission)

especially China, require vast quantities of elements present in sandalwood to cool nuclear reactors and machinery which are being used in reducing the radiation of nuclear reactors (Narasimhan and Gireesh Babu 2015) (Figs. 3.1 and 3.2).

Lumber pieces with the wavy grain margin are graded as "A" grade. A group of local artisans in Settigunta and Srikalahasti in Andhra Pradesh make toys, flower vases, water cups and other products from the material given to them by forest department, and products are sold through Lepakshi emporium (Figs. 3.3, 3.4, 3.5, 3.6, 3.7, 3.8 and 3.9).

However, the sources said there was a need to make value-added products from red sanders on a big scale as in China where the material was being used extensively for making furniture and interiors.

The Chinese use specially designed equipment for wood carving because of its toughness. The Chinese also believe that red sanders wood has strong aphrodisiac properties, and tablets made from a chemical extracted from red sanders were being sold in that country (Figs. 3.10, 3.11 and 3.12).

Fig. 3.2 Wood pieces of red sanders. (Courtesy Reddy 2018 with permission)

Fig. 3.3 Red sanders wood rosaries. (Courtesy Reddy 2018 with permission)

Fig. 3.4 Red sanders carved wood rosaries. (Courtesy Reddy 2018 with permission)

Fig. 3.5 Chess board coins made with red sanders wood. (Courtesy Reddy 2018 with permission)

3.2 Ethnobotany

Since time immemorial plants have been serving mankind in various ways, and man has been continuously using them for various requirements. A closer understanding of the association of man with the surrounding flora is essential for better utilization of plants in the service of mankind. Ethnobotany is defined as the study of the

Fig. 3.6 Rosaries made with red sanders wood. (Courtesy Reddy 2018 with permission)

Fig. 3.7 Carved wood plate made from the wood of red sanders. (Courtesy Reddy 2018 with permission)

association, interaction and interrelationships of human societies especially primitive human societies like tribal and aboriginal communities with the surrounding flora. Red sanders has attracted the attention of many of these ethnic communities for curing their ailments which are given below.

The heartwood of red sanders has various uses in traditional medicines and is popular for the treatment of diabetes apart from other ailments.

Herpes: Tribes of costal Karnataka administer red sanders as an anti-inflammatory agent for treating herpes (Bhandari and Chandrashekar 2011).

Ulcer treatment: Tribes of Yerukula and Irula of Chittoor district of Andhra Pradesh use the plant parts for curing ulcer (Vedavathy et al. 1997).

Fig. 3.8 Wood carving from the wood of red sanders. (Courtesy Redd 2018 with permission)

Fig. 3.9 Intricate wood carving from red sanders wood. (Courtesy Reddy 2018 with permission)

Jaundice: Powdered bark of stem is boiled in water till the volume is reduced to half the original content, and jaggery is added and made into pills, to administer in case of acute jaundice for 10 days. This is used by tribes of Shimoga region of Western Ghats for snake bite, diabetes and fever (Manjunatha 2006). Red sanders though does not occur in Shimoga district; perhaps it must have been planted there by forest department.

Blood purification: Wood paste is administered as blood purifier and remedy for skin diseases and poisonous affections in Malamalasar tribe of Kerala (Yeshodharan and Sujana 2007). Few tribal groups of Orissa like Kalahandi also use the wood as diuretic and blood purifier.

Fig. 3.10 Chinese wood carving using red sanders. (Courtesy Reddy 2018 with permission)

Fig. 3.11 Decorative wood carving. (Courtesy Reddy 2018 with permission)

Piles: The tribes of Kandhas in Kandhamal district of Odisha administer the decoction of *Calamus tenuis* roots and stem bark of *Azadirachta indica* and *Pterocarpus santalinus* with honey and sugar in an empty stomach for 21 days to treat piles (Behera et al. 2006).

Chest pain and tuberculosis: The tribes of Kalahandi, Orissa, administer the paste of the wood on the chest and orally consume the water with wood extract for tuberculosis and chest pain (Panda and Padhy 2008).

Fig. 3.12 Furniture made with red sanders wood. (Courtesy Reddy 2018 with permission)

Conjunctivitis: Sapwood extract of the plant is used for 3 days to treat conjunctivitis in the tribes of Chittagong district of Bangladesh and Tripura community of Hazarikhil (Faruque and Uddin 2011).

Skin infections: The powdered lumber wood is made into paste with egg and flour, to apply on the body for fair and smooth skin (Saikia et al. 2006). The powder made into a paste is also used as remedy for prickly heat. It is also used as cosmetic and curative for skin diseases by people of Assam (Saikia et al. 2006) (Fig. 3.13).

3.2.1 Pharmacognostic Studies

The importance of pharmacognosy has been widely felt in recent times. Pharmacognostic study includes parameters which help in identifying adulteration even in the dry powder form. This is again necessary because once the plant is dried and made into powder form, it loses its morphological identity and is easily prone to adulteration. Pharmacognostic studies ensure plant identity and lay down standardization parameters which will help and prevent adulterations. Such studies will help in authentication of the plants and ensure reproducible quality of herbal products which ensure safety and efficacy of natural products.

The macroscopic, microscopic, powder, organoleptic characters and fluorescence properties of the heartwood of *Pterocarpus santalinus* have been studied by Youngken (1948), Rao and Purkayastha (1972) and Donga et al. (2017).

Fig. 3.13 Use of red sanders powder in traditional medicine. (Courtesy Reddy 2018 with permission)

3.3 Microscopic Characteristics

Donga et al. (2017) gave the macroscopic and microscopic characters of the leaf and stem of *P. santalinus*, details of which are given below.

3.3.1 Transverse Section of Petiole

The transverse section of *P. santalinus* petiole shows the following features. Epidermis is single layered. Cortex consisted of eight to nine layers with white latex secretory gland. The vascular bundle "bean" or "pea" shaped varied from centre to margin. The vascular bundles are conjoint, collateral and open type (Donga et al. 2017).

3.3.2 Leaves

The transverse section of leaf shows the following features. The epidermis is single layered and palisade tissue is present below the epidermis. Vascular bundles are conjoint, collateral and open type placed on ventral surface. The cortex consisted of seven to eight layers of parenchymatous tissue and cluster crystals of calcium oxalate. The reticulate venation is seen on the lower surface of the leaf; anomocytic stomata are present on the lower epidermis (Donga et al. 2017).

3.3.3 Stem

The transverse section of *P. santalinus* stem shows the following. The epidermis is single layered with thick cuticle and many unicellular trichomes. The cork cambium is two to three layers and consists of cells with chloroplasts, and the pericycle surrounded the vascular bundles. The vascular bundles are in concentric type of arrangement; secondary xylem consists of metaxylem, xylem vessels, tracheids and fibrotracheid. Secondary phloem consists of sieve tubes, companion cells and medullary rays, and it is radially elongated. The cortex consists of seven to eight layers of polygonal parenchymatous cells. The pith is small and consists of parenchymatous cells with cluster of crystals of calcium oxalate (Donga et al. 2017).

3.3.4 Microscopy of Leaf Powder

The crude powder of leaves of *P. santalinus* leaves is dark green in colour and has characteristic odour and slight bitter taste. The specific characteristics determined from the leaf powder study under microscopic investigation showed border pitted vessels, anomocytic stomata, phloem, spiral and reticulate vessels.

3.3.5 Microscopy of Stem Powder

The crude powder of *P. santalinus* stem is reddish brown in colour and fine and has fragrant odour and aesthetic taste (rasa). The specific characteristics determined from the powder study under microscopic investigation showed annular vessels, border pitted vessels, pitted vessels and scalariform vessels.

The microscopic characteristics laid down are characteristics of the plant, and these parameters will be helpful in identifying the genuine drug especially its powder form.

3.3.6 Fluorescence Analysis

If the plant has a characteristic fluorescence, this can be an important tool to check the quality of the powdered material. Red sanders exhibit fluorescence under visible and UV light, and the same is summarized in Tables 3.1, 3.2 and 3.3.

Selvam (2012) has given identification protocols for red sanders heartwood. Some of the details are as given under.

Table 3.1 Fluorescence analysis of *Pterocarpus santalinus* leaf powder (After Donga et al. 2017)

Treatment	Visible light	Under UV light short wavelength (254 nm)	Long wavelength (365 nm)
1 N NaOH (aq)	Green	Black	Green
1 N NaOH (alc)	Green	Black	Green
Ammonia	Green	Black	Brown
Picric acid	Green	Black	Green
Petroleum ether	Green	Black	Green
50% HCl	Light green	Black	Brown
50% H_2SO_4	Green	Black	Green
Ethyl acetate	Green	Black	Brown
Ethyl alcohol	Green	Black	Brown
Methanol	Green	Black	Dark brown

Table 3.2 Fluorescence analysis of *Pterocarpus santalinus* stem powder (After Donga et al. 2017)

Treatment	Visible light	Under UV light short wavelength (254 nm)	Long wavelength (365 nm)
1 N NaOH (aq)	Green	Black	Green
1 N NaOH (alc)	Brown	Black	Green
Ammonia	Brown	Black	Green
Picric acid	Yellow	Black	Green
Petroleum ether	Brown	Black	Green
50% HCl	Brown	Black	Green
50% H_2SO_4	Brown	Black	Brown
Ethyl acetate	Brown	Black	Green
Ethyl alcohol	Brown	Black	Green
Methanol	Light Brown	Black	Light yellow

Table 3.3 Fluorescence analysis of *Pterocarpus santalinus* bark powder (After Donga et al. 2017)

Treatment	Visible light	Under UV light short wavelength (254 nm)	Long wavelength (365 nm)
1 N NaOH (aq)	Wine red	Black	Dark red
1 N NaOH (alc)	Wine red	Black	Green
Ammonia	Red	Black	Brown
Picric acid	Red	Black	Red
Petroleum ether	Red	Black	Red
50% HCl	Red	Black	Brown
50% H_2SO_4	Red	Black	Red
Ethyl acetate	Brown	Green	Yellow
Ethyl alcohol	Red	Black	Brown
Methanol	Reddish brown	Black	Brown

3.3.7 Exomorphic and Organoleptic Features of Heartwood

The scattered information available on the exomorphic (macro morphological) features and organoleptic characters of the heartwood of *P. santalinus* were collected and reviewed by Selvam (2012).

In transectional view of heartwood, growth rings are not evident. Vessels are diffuse in distribution, wide, thick-walled, circular or elliptical, and mostly solitary or less frequently in short radial multiples of two or three, often filled with dark gummy material. Vessels show bordered pits on the lateral walls. Diameter of the vessels ranges from 75 to 200 μm (Selvam 2012).

Axial xylem parenchyma paratracheal, banded, forming wing like lateral expansion or wavy tangential lines of three to five cells wide. Xylem fibres are thin-walled with wide lumen and lignified walls. Prismatic calcium oxalate crystals are occasionally seen within the fibres. In TLS view, xylem rays are thin, straight, storied and uniseriate.

3.3.8 Powder Microscopy

Under polarized light, the powdered heartwood of *P. santalinus* exhibits fibres, vessel elements, xylem (wood) parenchyma, starch grains and calcium oxalate crystals. Abundance of wood fibres is seen in the powdered preparation. They are narrow and elongated with tapering ends. The vessel elements are narrowly cylindrical and thick-walled with dense, circular or elliptical, wide, lateral wall pits. Non-lignified broken pieces of xylem parenchyma cells are often seen in the powder (Selvam 2012).

Two types of starch grains are seen very rarely in the powder, concentric starch grains with central hilum and plus (+)-shaped dark markings and elliptical starch grains with eccentric hilum and "V"- and "YY"-shaped dark lines/markings. The presence of starch grains is reported for the first time in this plant species. Prismatic calcium oxalate crystals of varying sizes and shapes are occasionally seen in the powder.

3.3.9 Maceration

The macerated heartwood of *P. santalinus* exhibits wood fibres, xylem (vessel) elements, ray parenchyma and crystals. Vessel elements are frequently seen in the macerated preparation, which are wide, short and cylindrical with vertical or horizontal parallel rows of pits. Xylem fibres are thick-walled with reduced lumen and lignified walls, which range from 480 to 950 μm in length. Fibres with ray parenchyma cells attached on the lateral walls are also seen. Prismatic calcium oxalate crystals are occasionally seen in the macerated preparation.

3.3.10 SEM Studies

Selvam (2012) carried out SEM studies on heartwood of *Pterocarpus santalinus*. A small piece of heartwood of *P. santalinus* from the peripheral region (last formed tissue) was taken up for the study. In cross-sectional view, it shows wide, thick-walled, circular or elliptical vessels in radial multiples of three. Some of the vessel elements are filled with gummy material. The diameter of the vessels ranges from 75 to 200 μm. Lateral wall pits of the vessels are elliptical, narrow and alternated, ranging from 3 to 11 μm in length. Xylem fibres are heavily thick-walled with reduced lumen. Xylem rays are narrow, straight, undulated and parallel to each other (Selvam 2012).

3.3.11 Fluorescence Analysis

The characteristic fluorescent properties (in the form of colours) emitted by the coarsely powdered heartwood of *P. santalinus* under ultraviolet radiation at 254 nm and the colours observed in visible light, before and after treating with various reagents, were recorded by Selvam (2012).

The powdered heartwood of this plant as such appeared reddish orange in visible light and deep reddish orange under ultraviolet radiation. After treating with various reagents, in visible light, it showed different shades of yellow in most of the cases except hexane, methanol, acetone, aqueous and alcoholic 1N sodium hydroxide extracts. The hexane extract did not show any colour. However, the methanol and acetone extracts revealed deep brownish red and deep orange colours, respectively. Though the water extract did not show any solid colour (appears almost colourless), a very pale brown colour with light yellow tinge was observed. The aqueous and alcoholic 1N sodium hydroxide extracts showed reddish brown colours. Under ultraviolet radiation, chloroform, acetone and 50% sulphuric acid and 50% nitric acid-treated extracts fluoresced different shades of yellow. The water and methanol extracts showed pale green and magenta or deep orange colours, respectively. The aqueous and alcoholic 1N sodium hydroxide extracts revealed dark violet colour, but the hexane- and 1N hydrochloric acid-treated extracts fluoresced no colours.

Selvam (2012) gave the exomorphic features and organoleptic characters of the heartwood of *Pterocarpus santalinus*, the details of which are given below. Surface is hard with mild ridges and furrows at different angles with longitudinal striations. Transversely cut surface shows darker and lighter zones with varying colours, both internally and externally, viz. deep red or reddish brown or dusky red to dark reddish brown or black. It has no specific odour (inodorous) except mild woody smell. Taste is slightly astringent and bitter. The heartwood is very hard but can be easily split longitudinally. Fracture is very hard and splintery. Fractured surface is uneven, fibrous or horny. Heartwood shows apparently visible deep reddish brown colour in its crude form (cut pieces and coarse powder), which is one of the constant characters, helpful in spot identification (Table 3.4).

Table 3.4 Fluorescence properties of the heartwood of *Pterocarpus santalinus* under visible light vis-à-vis ultraviolet radiation at 254 nm (After Selvam 2012)

S. No.	Tests	Visible light	UV radiation
1	Plant powder	Reddish	Deep reddish orange
2	Plant powder treated with water	Colourless (or) very pale brown with pale green	Light yellow tinge
3	Plant powder treated with hexane	Colourless	Colourless
4	Plant powder treated with chloroform	Lemon yellow	Pale lemon yellow
5	Plant powder treated with methanol	Deep brownish red	Magenta or deep orange
6	Plant powder treated with acetone	Deep orange	Deep chrome yellow
7	Plant powder treated with 1N NaOH in water	Reddish brown	Dark violet
8	Plant powder treated with 1N NaOH in methanol	Reddish brown	Dark violet
9	Plant powder treated with 1N hydrochloric acid	Pale yellow	Colourless
10	Plant powder treated with 50% sulphuric acid	Chrome yellow	Light ochre yellow
11	Plant powder treated with 50% nitric acid	Golden yellow	Light ochre yellow

1 g of plant powder (heartwood) treated with 10 ml of reagent solution. The tests pertaining to the serial numbers 4, 5, 6, 7, 8, 10 and 11 showed immediate reactions by yielding colours that were observed in daylight

Table 3.5 Specific microscopic characters observed in the heartwood of *Pterocarpus santalinus* (Wambatuwewa 2009)

Character	*Pterocarpus santalinus* heartwood
Distribution of vessels	Diffuse porous
Arrangement of vessels	Mostly isolated, rarely multiples
Deposits in vessels	Red in colour, mostly filled
Pits on vessel walls	Bordered
Arrangement of axial parenchyma	Aliform or confluent
Ray parenchyma	Uniseriate
Type of wood	Storied
Number of vessels/microscopic field (10 × 10)	11 ± 3
Average diameter of vessels (10 × !0)	0.39 ± 0.2
Fibres: vessel ratio	1750: 1
Crystals in parenchyma	Present

Wambatuwewa (2009) gave the pharmacognostic characters of heartwood of *Pterocarpus santalinus* (Table 3.5). The specific gravity of *Pterocarpus santalinus is* 0.956 ± 25.15 (Wambatuwewa 2009).

3.4 Physicochemical Properties of Wood

In *Pterocarpus santalinus* extract, the moisture content of wood is 40%, total ash content in wood is 7, acid insoluble ash content is 0.3, water soluble ash content is 6.4%, alcohol soluble extract is 23.82% and total ash is 2% (Pandey et al. 2014).

3.5 Conclusion

As wood and powder are highly valued they are adulterated many times with certain other tree species resembling same colour, hence the study of crude woodchips and powder is helpful for scientific validation when used is preparation of drugs and other food products to avoid deleterious effects.

References

Behera SK, Panda A, Behera SK, Misra MK (2006) Medicinal plants used by the Kandhas of Kandhamal district of Orissa. Indian J Tradit Knowl 5:519–528

Bhandari MJ, Chandrashekar KR (2011) Herbal therapy of herpes in the ethnomedicine of coastal Karnataka. Indian J Tradit Knowl 10(3):528–532

Donga S, Moteriya P, Pande J, Chanda S (2017) Development of quality control parameters for the standardization of *Pterocarpus santalinus* Linn. f. leaf, stem. J Pharmacogn Phytochem 6 (4):242–252

Faruque O, Uddin SB (2011) Ethnodiversity of medicinal plants used by Tripura community of Hazarikhil in Chittagong district of Bangladesh. J Taxon Biodivers Res 5:27–32

Kumar ANA, Joshi G (2014) *Pterocarpus santalinus* (Red Sanders) an endemic, endangered tree of India: current status, improvement and the future. J Trop For Environ 4(02):1–10

Mahammadh RV (2014) Smuggling of red sandal in India: a review study. Int J Econ Bus Rev 2 (10):55–63

Manjunatha BK (2006) Hepatoprotective activity of *Pterocarpus santalinus* L. f., an endangered plant. Indian J Pharmacol 38:25–28

Narasimhan TE, Babu G Chennai. Last Updated at April 10, 2015 14:54 IST, Business behind red sandalwood: the endangered plant species. https://www.business-standard.com/article/current-affairs/business-behind-red-sandalwood-the-endangered-plant-species-115041000424_1.html

Panda T, Padhy RN (2008) Ethnomedicinal plants used by tribes of Kalahandi district, Orissa. Indian J Tradit Knowl 7:242–249

Pandey B, Gangrale D, Upadhyay N, Tiwari P (2014) Physicochemical analysis of *Pterocarpus santalinus* L. extracts. Indian J Sci Res 4(1):201–204

Ramabrahmam V, Sujatha M (2016) Red Sanders in Rayalaseema Region of Andhra Pradesh: importance to commercial & medicinal value. IOSR J Pharm Biol Sci 11(1):57–60

Rao KR, Purkayastha SK (1972) Indian woods, their identification, properties and uses. Volume 3. Leguminosae to Combretaceae. Manager, Publications, Delhi. 262 p

Reddy OSK (2018) Red Sanders – *Pterocarpus santalinus* – Rakta chandan – The Red Gold – Green wealth – an opportunity for Indian Farming community – let's explore. https://www.linkedin.com/pulse/red-sanders-gold-green-wealth-future-generations-indian-osk-reddy

Saikia AP, Ryakala VK, Sharma P, Goswami P, Bora U (2006) Ethnobotany of medicinal plants used by Assamese people for various skin ailments and cosmetics. J Ethnopharmacol 106:149–157

Selvam ABD (2012) Identification protocols for Red Sanders. In: Pharmacognosy of negative listed plants. Botanical Survey of India, Kolkata. pp 171–181

Vedavathy S, Sudhakar A, Mrudula V (1997) Tribal medicinal plants of Chittoor. Ancient Science of Life 26:307–331

Wambatuwewa WMKB (2009) Pharmacognostic studies on *Pterocarpus santalinus* L.f. and *Evolvulus alsinoides* (L.) L. M.Phil. Thesis, University of Sri Jayewardenapura, Sri Lanka

Wealth of India (1969). A dictionary of Indian raw materials and industrial products. NISCAIR, New Delhi

Wenbin H, Xiufang S (2013) Tropical hardwood flows in China: case studies of rosewood and Okoumé. Forest Trends Association, Washington, DC

Yeshodharan K, Sujana KA (2007) Ethnomedical knowledge among Malamalasar tribe of Parambikulam wildlife sanctuary, Kerala. Indian J Tradit Knowl 6(3):481–485

Youngken HW (1948) Text book of pharmacognosy. The Blackinston Company, Philadelphia

Chemistry of Red Sanders

4

T. Pullaiah, V. Damodara Reddy, Umalatha, and B. S. Rashmi

Abstract

Plants are commendable source of metabolites with therapeutic properties from time immemorial. *Pterocarpus santalinus*, an endemic Indian plant, is having an array of phytochemicals with a plethora of applications in medicine and industries. Phytochemical investigations and bioprospecting of leaf, stem, bark, and heartwood of *P. santalinus*, by employing various extraction methods, revealed the presence of different groups of metabolites like triterpenoids, alkaloids, flavonoids, phenols, saponins, glycosides, sterols, tannins, and more specific metabolites like pterocarpol; pterocarptriol; santalins A, B, and Y; isopterocarpalone; pterocarpodiolones; β-eudesmol; and cryptomeridiol of high medicinal value with significant applications in pharma, cosmetics, liquor, and textile industries. Here different classes of metabolites found in various parts of *Pterocarpus santalinus* are discussed, and a comprehensive review is presented.

Keywords

Phytochemicals · Santalin · Pterocarpol · Flavonoids · Glycosides · Phenols

T. Pullaiah (✉)
Department of Botany, Sri Krishnadevaraya University, Anantapur, Andhra Pradesh, India

V. Damodara Reddy
Department of Biochemistry, Sri Krishnadevaraya University, Anantapur, Andhra Pradesh, India

Umalatha · B. S. Rashmi
Padmashree Institute of Management and Sciences, Bangalore, Karnataka, India

T. Pullaiah et al. (eds.), *Red Sanders: Silviculture and Conservation*,
https://doi.org/10.1007/978-981-13-7627-6_4

4.1 Introduction

Plants are being extensively used as virtuous source of natural bioactive compounds worldwide since beginning of the mankind, in all the civilizations for their valuable medicinal uses and health benefits. Nowadays, this ancient practice has fetched the attention of scientific community in exploring the principles of traditional medicines for the discovery of novel drugs to treat various diseases and disorders. In the twenty-first century, with the concept of globe as a village, the modern technology, medicinal inventions are being made available to everyone. However, unfortunately, there are some areas where the accessibility to new drugs is less due to socioeconomic causes. In this regard, World Health Organization (WHO) has endorsed the evaluation of potential benefits of plants as effective therapeutic agents (Halim et al. 2011). Approximately, more than 80% of the drugs used in developed and fast developing countries are derivatives of phytochemicals (Halim and Misra 2011; Ramawat and Goyal 2008; Kumar et al. 2009).

India has a heritage of various medicinal plants in its rich biodiversity. *Pterocarpus santalinus* is one such plant with rich phytochemicals of highly therapeutic significance. Phytochemical analysis of *P. santalinus* revealed the presence of flavonoids, terpenoids, phenolic compounds, alkaloids, saponins, tannins, and glycosides. In addition nonspecific metabolites such as isoflavones, isoflavonoid glucosides, triterpenes, sesquiterpenes, and related phenolics have also been reported. Red sanders have specific phytoconstituents such as β-sitosterol, lupeol, epicatechin, lignans, and pterostilbenes (Kumar and Seshadri 1975; Krishnaveni and Rao 2000a, b). It also contains highly specific metabolites pterocarpol; pterocarptriol; santalins A, B, and Y; isopterocarpalone; pterocarpodiolones; β-eudesmol; and cryptomeridiol with high medicinal value (Yoganarasimhan 2000), and some of the specific metabolites of *P. santalinus* are depicted in Fig. 4.1.

The important phytochemical constituents of different parts of *P. santalinus* (bark, leaf, and heartwood) have been described below and listed in Table 4.1.

4.2 Bark

Phytochemical analysis of active fraction from bark of *P. santalinus* has mainly showed the presence of flavonoids, glycosides, and phenols. Biological testing of these bioactive compounds in animal models exhibited a significant antidiabetic activity by reducing the glycosylated haemoglobin and blood glucose levels, improving hyperlipidaemia, and restoring the insulin levels. This antidiabetic activity has been attributed to the improved levels of hepatic carbohydrate-metabolizing enzymes after the administration of active components of the bark (Kondeti et al. 2010). In addition, ethanolic extract of bark also showed the presence of an array of amino acids including aspartic acid, glutamic acid, alanine, serine, glycine, threonine, lysine, asparagine, histidine, arginine, cystine, glutamine, proline, phenylalanine, and leucine (Devi and Basha 2012), and it has been reported that amino acid composition influences the wavy or straight grain formation in *P. santalinus* (Das and Dayanand 1983, 1984).

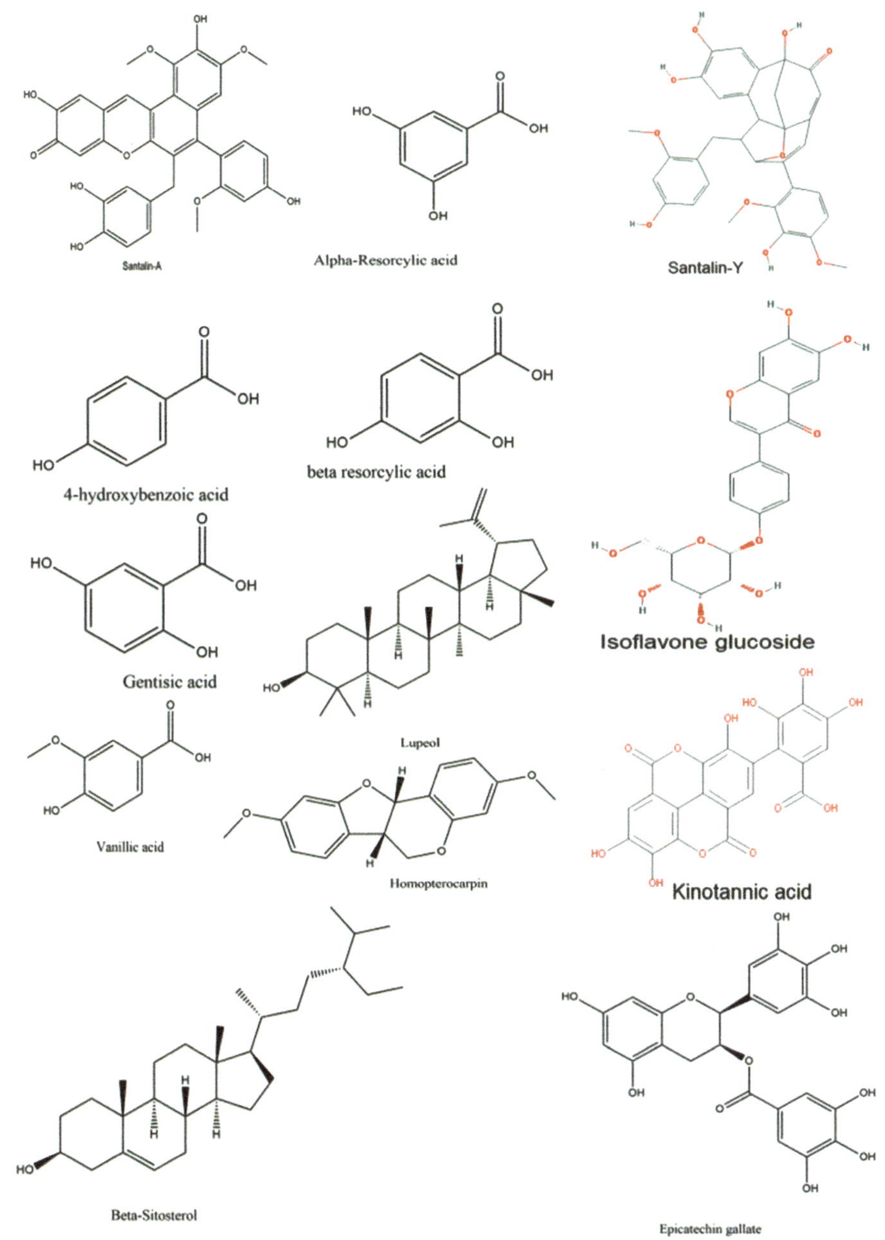

Fig. 4.1 Structures of some phytochemicals isolated from plant parts of *Pterocarpus santalinus*

Table 4.1 List of compounds isolated from *Pterocarpus santalinus*

S. No.	Name of compound	Plant part	References
Triterpene			
1	Lupenone, β-amyrone, epilupeol, β-amyrin, lupeol, stigmasterol, β-sitosterol	Leaves	Kumar and Seshadri (1975, 1976a)
2	Erythrodiol	Sapwood	Kumar et al. (1974)
3	Betulin	Stem, callus	Krishnaveni and Rao (2000a, c)
4	3-Ketooleanane	Bark	Kameswara Rao et al. (2001) and Kumar and Seshadri (1976b)
5	Lupenediol	Bark	Kumar and Seshadri (1975)
6	Lupenone, lup-(20)29-ene-2α, 3α-diol	Heartwood	Kumar et al. (1974)
7	Acetyloleanolic aldehyde, acetyloleanolic acid	Sapwood	Kumar et al. (1974)
Sesquiterpenoid			
1	Eudesmol; α-, β-, and γ-isomer; β-santalol	Heartwood	Kukla et al. (1976) and Kumar et al. (1974)
2	Pterocarpol	Heartwood	Kukla et al. (1976) and Kumar et al. (1974)
3	Isopterocarpolone, pterocarptriol, pterocarpdiolone	Heartwood	Kumar et al. (1974)
4	Cryptomeridiol	Heartwood	Kumar et al. (1974)
5	2,11-Dihydroxy-4(15)-eudesmene	Bark	Bahl et al. (1968)
6	β-Eudesmol		
7	Canusenol K, Canusenol L, Hamahasal A	Heartwood	Li et al. (2017)
	12,15-Dihydroxy curcumene		
	ent-4(15)-eudesmen-1α,11-diol		
	(3β)-eudesmen-4(14)-ene-3,11-diol		
Isoflavone			
1	Hydroxy-2′,4′,5′,7-tetramethoxy-isoflavone	Heartwood	Krishnaveni and Rao (2000a)
2	Neoflavones I and II	Stem	Hakamata et al. (1993)
3	Isoliquiritigenin (2,4,4′-trihydroxychalcone)	Stem	Sawhney and Seshadri (1956) and Krishnaveni and Rao (2000a)
4	Liquiritigenin (7,4′-dihydroxyflavone)	Stem	Raju and Raju (2000) and Kinjo et al. (1995a, b)
5	Marsupsin, pterosupin, liquiritigenin	Heartwood	Krishnaveni and Rao (2000a)
Lignans			
1	Savinin	Heartwood	Cho et al. (2001)
2	Calocedrin	Heartwood	Cho et al. (2001)

(continued)

Table 4.1 (continued)

S. No.	Name of compound	Plant part	References
Coumarin			
1	Yellow 3-arylcoumarin derivative of santalin A	Stem	Kinjo et al. (1995a, b)
2	5-Hydroxy-7-O-(3-methyl)-but-2-enylcoumarin	Heartwood	Singh et al. (1993)
3	6-Hydroxy-7-methoxy-2H-chromen-2-one	Heartwood	Azamthulla et al. (2016)
Stilbenes			
1	Pterostilbene	Heartwood	Charvet-Faury et al. (1998), King et al. (1953), and Sawhney and Seshadri 1956
Phenolics			
1	Melilotic acid, esculetin, quercetin-3-rutinoside	Seed	Venkataramaiah et al. (1980)
Essential oil			
1	Darse	Heartwood	Shankaranarayana et al. (1993)
Metals in red sanders			
1	Uranium, thorium	Heartwood	Raju and Raju (2000)
2	Zinc, copper, strontium	Heartwood	Raju and Nagaraju (1999)
Glycosides			
1	4′,5-Dihydroxy-7-O-methyl-isoflavone-3′-O-β-D-glucoside	Heartwood	Krishnaveni and Rao (2000b, d)
2	4′,5-Dihydroxy-7-O-methyl-isoflavone-3′-O-β-D-(3″-E-cinnamoyl)-glucoside	Heartwood	Singh et al. (1992)
3	7-Hydroxy-6-methoxy-coumarin-7-O-β-D-apiofuranosyl-(1→6)- β-D-glucopyranoside	Heartwood	Singh et al. (1993)
4	7-Hydroxy-6-methoxy-coumarin-7-O-α-L-arabinopyranosyl-(1→3)- β-D-galactopyranosyl-(l-→6)- β-D-galactopyranoside	Heartwood	Kesari et al. (2004)
5	6-Hydroxy-5-methyl-3′,4′,5′-trimethoxy-aurone-4-I-α-L-rhmnopyranoside	Wood	Kesari et al. (2004)
6	6,4′-Dihydroxy-aurone-4-I-rutinoside	Wood	
Benzofurans and neoflavonoids			
1	Pterolinuses A–J	Heartwood	Wu et al. (2011a, b)
2	Melanoxin		
3	Melanoxoin		

(continued)

Table 4.1 (continued)

S. No.	Name of compound	Plant part	References
4	S-3-Hydroxy-4,4-dimethoxydalbergione		
5	Melanin		
Miscellaneous compounds			
1	Santalin, santal	Heartwood	Krishnaveni and Rao (2000a, c), Patrick et al. (1998), Tennekone et al. (1998), Arnone et al. (1972, 19750) and Gurudutt and Seshadri (1974)
2	Santalin A	Stem	Kinjo et al. (1995a, b), Kumar et al. (1974), Ravindranath and Seshadri (1972, 1973) and Raju and Nagaraju (1999)
3	Santalin Y	Whole plant	Kinjo et al. (1995a, b), Arnone et al. (1975), Kumar et al. (1974), and Ravindranath and Seshadri (1973)
		Sapwood	
4	(-)-Fistenidol, santarubin A, santarubin B	Stem	
5	Santalin B		
6	Pterocarpin, homopterocarpin	Bark	King et al. (1953)
Hydroxybenzoic acids			
7	Para-hydroxybenzoic acid, gentisic acid, α-resorcylic acid, β-resorcylic acid, vanillic acid	Heartwood	Khadem and Marles. (2010)
Phenanthrenedione			
8	Pterolinus K	Heartwood	Cai et al. (2004)

4.3 Leaf

Phytochemical analysis of leaves of *P. santalinus* showed the presence of steroids, flavonoids, carbohydrates, and mainly triterpenes: lupenone, β-amyrone, epilupeol, β-amyrin, lupeol, stigmasterol, and β-sitosterol. These triterpenes confer anti-inflammatory, analgesic, antioxidant, and antimicrobial effects (Bishayee et al. 2011; Arokiyaraj et al. 2008). Therefore, decoction of red sandalwood leaves has been used in alleviating the burden of gout, rheumatism, and other inflammatory diseases in traditional medicinal practice.

4.4 Stem and Heartwood

Phytochemical analysis of *P. santalinus* heartwood showed the presence of carbohydrates, flavonoids, terpenoids, phenolic compounds, alkaloids, saponins, tannins, and glycosides. *Pterocarpus* species are found to be rich in isoflavonoids, terpenoids, as well as related phenolic compounds, β-sitosterol, lupeol, and epicatechin (Kesari et al. 2004). As mentioned above, *P. santalinus* also contains several highly specific metabolites such as pterocarpol; pterocarptriol; santalins A, B, and Y; isopterocarpalone; pterocarpodiolones; β-eudesmol; and cryptomeridiol (Yoganarasimhan 2000). Santalin extracts are used in colouring pharmaceutical preparations, foodstuffs, high-class alcoholic liquors, paper, pulp, textiles, and wine. They are also used in manufacturing of dyes, plastics, pesticides, explosives, and drugs (Cai et al. 2004; Krishnaveni and Rao 2000a; Patrick et al. 1998; Tennekone et al. 1998).

Stilbenes are secondary plant metabolites synthesized from phenylpropanoid pathways. Stilbenes are phenolic compounds and exist in two stereoisomer forms, namely, E-stilbene (*trans*) and Z-stilbene (*cis*). Interestingly, a specific stilbene, namely, pterostilbene methyl ether, of resveratrol was reported to be found in *P. santalinus* heartwood and fetched attention of pharma sector for exhibiting promising anti-inflammatory, antioxidant, antineoplastic, and anticancer properties (Giacomini et al. 2016; Charvet-Faury et al. 1998; Sawhney and Seshadri 1956; King et al. 1953). In addition, pterostilbene is also known to act as phytoalexin in plants and protects wood from plant pathogens by exhibiting broad spectrum antimicrobial activity (Langcake and Pryce 1977).

Another class of important plant metabolite are lignins and lignans. These compounds deposit on the secondary cell wall in plants and play an important role in defence against pathogens (Bagniewska-Zadworna et al. 2014). Lignans also exhibit significant role in plants' defence mechanism against insects by regulating insect feeding and altering its physiology (Harmatha and Dinan 2003). Further, biosynthesis of lignins in plants extensively contributes to development of organ/tissue, plant growth, mounting resistance, and inducing specific response to various abiotic and biotic stresses (Liu et al. 2018). Savinin, calocedrin, and eudesmin were some of the specific lignans reported to be present in the heartwood of *P. santalinus* which majorly contribute to the disease and insect resistance traits of the plant (Cho et al. 2001). Lignins are found to exhibit many therapeutic properties. They exhibit potential anti-inflammatory or antioxidant activity. In addition, they are proven as mitotic toxins, and therefore their derivatives are in use as chemotherapeutic agents.

Further, isoflavones are group of plant secondary metabolites mainly found in *P. santalinus* and other members of Fabaceae. Isoflavones play an important role in both antagonistic and mutualistic plant-microbe interactions. In case of mutualistic interaction, isoflavones act as chemoattractants for root nodule-forming rhizobia and also as an inducer for expression of *nod* gene, whereas in case of antagonistic interaction, isoflavones participate in disease resistance response (Van Rhijn and Vanderleyden 1995; Pueppke 1996). They exhibit estrogenic and antiestrogenic effects depending on the target tissue. They are well known for antioxidant, antimicrobial, and anti-inflammatory health benefits. They also possess antioxidant and anticancer activities. They play a crucial role in preventing hormone-related cancers like breast cancer, cervical cancer, and the male prostate or testicular cancer (Krishnaveni and Rao 2000a; Raju and Raju 2000). Hydroxy-2′,4′,5′,7-tetramethoxy-isoflavone; neoflavones I and II; isoliquiritigenin (2,4,4-′-trihydroxychalcone); and liquiritigenin (7,4′-dihydroxyflavone) are some of the isoflavones found in *P. santalinus* (Yu et al. 2016; Kurzer 2000). The heartwood contains isoflavone glucosides and two antitumour lignans, viz. savinin and calocedrin.

Triterpenes, a class of compounds consisting three terpene units, are one of the largest and diverse groups of plant metabolites. Simple forms of triterpenes are found in surface waxes as well as in specialized membranes and mainly act as signalling molecules, while the complex glycosylated triterpenes confer protection against pathogens and pests (Thimmappa et al. 2014; Kumar et al. 1974). Lupenone, β-amyrone, epilupeol, β-amyrin, lupeol, stigmasterol, β-sitosterol, erythrodiol, acetyloleanolic aldehyde, acetyloleanolic acid, 3-ketooleanane, and lupenediol are some of the important triterpenes reported to be found in *P. santalinus* (Kameswara Rao et al. 2001; Kumar and Seshadri 1975, 1976b; Kumar et al. 1974). These metabolites have anti-inflammatory, analgesic, antipyretic, hepatoprotective, cardiotonic, sedative, and tonic effects. They are known to confer antioxidant, antiviral, antimicrobial, antiallergic, antipruritic, and antiangiogenic spasmolytic activity. Apart from that they exhibit cytotoxicity against a variety of cancer cells without manifesting any toxicity in normal cells (Kameswara Rao et al. 2001; Kumar and Seshadri 1975, 1976a; Kumar et al. 1974). Further, sesquiterpenes are another class of terpenes consisting of three isoprene units, and structurally they may be acyclic or contain rings. Biochemical modifications such as oxidation or rearrangement of functional group lead to the formation of related compounds called sesquiterpenoids (Breitmaier 2006). Sesquiterpenes in plants act as semiochemicals (defensive agents or pheromones) and participate in defence action. Eudesmol; α-, β- and γ-isomer; β-santalol, pterocarpol, isopterocarpolone, pterocarptriol, pterocarpdiolone, cryptomeridiol, canusenol K, canusenol L, hamahasal A, 12,15-dihydroxy curcumene, *ent*-4(15)-eudesmen-1α, and 11-diol (3β)-eudesm-4(14)-ene-3,11-diol are some of the sesquiterpenes found in *P. santalinus* (Li et al. 2017; Ivanescu et al. 2015; Kumar et al. 1974).

Coumarins are ubiquitous secondary metabolites of higher plants majorly responsible for defence mechanism against pathogens, response to abiotic stress, regulation of phytohormones, and oxidative stress (Bourgaud et al. 2006; Vuky and Motzer 2000). Coumarins act as anticoagulant drugs. They block regeneration and recycling of vitamin K. They are useful in the treatment of high-protein oedema (HPE) and other chronic infections. As coumarins are immunostimulatory drugs, they are also used in the treatment of malaria. Coumarins can be used not only to treat cancer but also to treat the side effects caused by radiotherapy (Kinjo et al. 1995a, b). The coumarins reported in *P. santalinus* include 5-hydroxy-7-O-(3-methyl)-but-2-enylcoumarin, 6-hydroxy-7-methoxy-2H-chromen-2-one, and 3-aryl coumarin (Azamthulla et al. 2016; Kinjo et al. 1995a, b; Singh et al. 1993).

In addition, glucosides, viz. 4′,5-dihydroxy-7-O-methyl-isoflavone-3′-O-β-D-glucoside, 4′,5-dihydroxy-7-O-methyl-isoflavone-3′-O-β-D-(3″-E-cinnamoyl)-glucoside, 7-hydroxy-6-methoxy-coumarin-7-O-β-D-apiofuranosyl-(1→6)- β-D-glucopyranoside, and 7-hydroxy-6-methoxy-coumarin-7-O-α-L-arabinopyranosyl-(1→3)- β-D-galactopyranosyl-(l-→6)- β-D-galactopyranoside, are also present in *P. santalinus*. Glucosides in general are mutagenic, and they can cause intestinal tumours. They are potential prodrugs. Toxicity is mostly dose-related, and it may lead to gastrointestinal disturbances, neurological disturbances, distinctive changes on the electrocardiogram (ECG), and consequences of intracellular Ca^{2+} overload, gynaecomastia, and increased vagal activity. Glucosides also stimulate cardiac muscle contractions (Krishnaveni and Rao 2000b; Singh et al. 1992, 1993).

Further, 30 naturally occurring phenolic acids, particularly hydroxyl and polyhydroxy benzoic acids, have been reported to have biological activities; among them 3-hydroxybenzoic acid, gentisic acid, α- and β-resorcylic acid, and vanillic acid are of great significance in pharma industries (Khadem and Marles 2010). These hydroxybenzoic acids are peroxynitrite scavengers, thus having antioxidant property.

Comparison of nuclear magnetic resonance and mass spectrometry data identified the presence of dehydromelanoxin-6, melanoxin-7, melanoxoin-14, S-30-hydroxy-4, 40-dimethoxydalbergione-15, and melannein-16 in the heartwood of *P. santalinus* (Wu et al. 2011a, b). These benzofurans and neoflavonoids have cytotoxic and anti-inflammatory activities.

A comprehensive list of phytochemicals reported by Bisby (1994) **in *Pterocarpus santalinus.***

Simple Aromatic Natural Products
2,4-Dihydroxybenzoic acid (various parts)
3,5-Dihydroxybenzoic acid (various parts)
Ferulic acid (*E-form*) (various parts)
4-Hydroxy-3-methoxy-benzoic acid (various parts)
3-(4-Hydroxyphenyl)-2-propeonic acid; *(E)-form* (various parts)
Pterostilbene (heartwood)
Syringic acid (various parts)

Benzopyranoids
2H-1-Benzopyran-2-one (various parts)

Flavonoids
Homopterocarpin (wood)
Pterocarpin (wood)
Santalin A (heartwood, sapwood)
Santalin B (heartwood, sapwood)
Santalin C (heartwood)

Terpenoids:
Acetyl oleanolic aldehyde (sapwood)
β-Amyrin (leaf)
β-Amyrone (bark, leaf)
Betulin (bark, leaf)
Cedrol (wood)
Cryptomeridiol (heartwood)
Epilupeol (bark, leaf)
Erythrodiol (leaf, sapwood)
β-Eudesmol (heartwood)
Isopterocarpalone (heartwood)
20(29)-Lupene-2-3-diol; (2α,3α)-form (bark)
20(29)-Lupene-2-3-diol; (2α,3 β)-form (bark)
Lupenone (bark, leaf)
Lupeol (bark, leaf)
Oleanolic acid: Ac (heartwood, sapwood)
Pterocarpdiolone (heartwood)
Pterocarpol (heartwood, sapwood)
Pterocarptriol (heartwood)

Steroids
β-Sitosteerol (bark, leaf)
Stigmasterol (leaf)

P. santalinus is a plant with large number of diverse phytochemicals which are highly complex compounds. The compounds identified so far are general preliminary compounds, and further research is required to explore more on these phytochemicals.

References

Arnone A, Merlini L, Nasini G (1972) Santalin pigments on the structure of santalin A. Tetrahedron Lett 33:3503–3506

Arnone A, Camarda L, Merlini L, Nasini G (1975) Structure of the red sandalwood pigments santalin A and santalin B. J Chem Soc Perkin Trans 1:186–194

Arokiyaraj S, Radha R, Martin S, Perinbam K (2008) Phytochemical analysis and anti-diabetic activity of *Cadaba fruticosa* R. Br Indian J Sci Technol 1(6):1–4. Domain site: http://www.indjst.org

Azamthulla M, Anbu J, Babu VLA, Rajkapoor B (2016) Isolation and characterization of *Pterocarpus santalinus* heartwood extract. Pharm Lett 8(12):34–39

Bagniewska-Zadworna A, Arasimowicz-Jelonek M, Smoliński DJ, Stelmasik A (2014) New insights into pioneer root xylem development: evidence obtained from *Populus trichocarpa* plants grown under field conditions. Ann Bot 113(7):1235–1247

Bahl CP, Parthasarathy MR, Seshadri TR (1968) Constitution of Pterocarpol. Tetrahedron 24:6231–6235

Bisby F (1994) Phytochemical dictionary of the Leguminosae, vol 1. Chapman & Hall, London

Bishayee A, Ahmed S, Brankov N, Perloff M (2011) Triterpenoids as potential agents for the chemoprevention and therapy of breast cancer. Front Biosci 16:980–996
Bourgaud F, Hehn A, Larbat R, Doerper S, Gontier E, Kellner S, Matern U (2006) Biosynthesis of coumarins in plants: a major pathway still to be unravelled for cytochrome P 450 enzymes. Phytochem Rev 5:293–308. https://doi.org/10.1007/s11101-006-9040-2
Breitmaier E (2006) Terpenes—flavors, fragrances, pharmaca, pheromones. Wiley VCH, Weinheim
Cai YZ, Luo Q, Sun M, Corke H (2004) Antioxidant activity and phenolic compounds of 112 traditional Chinese medicinal plants associated with anticancer. Life Sci 74:2157–2184
Charvet-Faury S, Derbesy M, Sochini F, Derbesy F (1998) Sandalwood extract (*Pterocarpus santalinus*). Study of antioxidant and anti-UV effects. Riv Ital EPPOS (Spec Num) 435–458
Cho JY, Park J, Kim PS, Yoo ES, Bik KV, Park MH (2001) Savinin, a lignin from *Pterocarpus santalinus* inhibits tumor necrosis factor – alpha production and T cell proliferation. Biol Pharm Bull 24(2):167–171
Das TL, Dayanand T (1983) Amino acid composition in wavy-grained and straight grained trees of Red Sanders (*Pterocarpus santalinus* Linn. f.). Indian J For 6:158–159
Das TL, Dayanand T (1984) Some distinguishing characteristics between wavy grained and straight grained trees of red Sanders *Pterocarpus santalinus*. Indian J For 7:69–71
Devi RKSL, Basha SKM (2012) Comparative amino acid profiling of *Pterocarpus santalinus* and *Boswellia ovalifoliolata* barks of Tirumala Hills, Eastern Ghats, Andhra Pradesh. Int J Basic Appl Chem Sci 2(4):102–106
Giacomini AC, Abreu MS, Giacomini LV, Siebel AM, Zimerman FF, Rambo CL, Mocelin R, Bonan CD, Piato AL, Barcellos LJ (2016) Fluoxetine and diazepam acutely modulate stress induced-behavior. Behav Brain Res 296:301–310
Gurudutt KN, Seshadri TR (1974) Constitution of the pigments santalin A and B. Phytochemistry 13:2845–2848
Hakamata J, Hiarayama Y, Itsukida T (1993) Antiandrogenic neoflavones from red sandal wood. Jpn Kokai Tokkyo Koho. 7pp
Halim ME, Misra A (2011) The effect of aqueous extract of *Pterocarpus santalinus* heartwood and vitamin E supplementation in streptozotocin-induced diabetic rats. J Med Plants Res 5:398–409
Halim R, Gladman B, Danquah MK, Webley PA (2011) Oil extraction from microalgae for biodiesel production. Bioresour Technol 102(1):178–185
Harmatha J, Dinan L (2003) Biological activities of lignans and stilbenoids associated with plant-insect chemical interactions. Phytochem Rev 2:321–330
Ivanescu B, Miron A, Corciova A (2015) Sesquiterpene lactones from *Artemisia* genus: biological activities and methods of analysis. J Anal Meth Chem. Article ID 247685, 21 pages. http://dx.doi.org/https://doi.org/10.1155/2015/247685
Kameswara Rao B, Giri R, Kesavulu MM, Apparao C (2001) Effect of oral administration of bark extracts of *Pterocarpus santalinus* L. on blood glucose level in experimental animals. J Ethnopharmacol 74(1):69–74
Kesari AN, Gupta RK, Watal G (2004) Two aurone glycosides from heartwood of *Pterocarpus santalinus*. Phytochemistry 65(23):3125–3129
Khadem S, Marles RJ (2010) Monocyclic phenolic acids; hydroxyl and polyhydroxybenzoic acids: occurrence and recent bioactivity studies. Molecules 15:7985–8005
King FE, Cotterill CB, Godson DH, Jurd L, King TJ (1953) The chemistry of extractives from heartwoods XIII. Colourless constituents of *Pterocarpus* spp. J Chem Soc 75:3693–3697
Kinjo J, Uemura H, Nohara T, Ono M, Ito Y (1995a) Novel santalin analogs from *Pterocarpus santalinus* (Leguminosae) their biogenesis and anti-oxidative activities. Yoshihira K. Jpn Tennen Yuki Kagobutsu Toronkai Koen Yoshishu 37:493–498
Kinjo J, Uemura H, Nohara T, Yamashita M, Marubayashi N, Yoshihara K (1995b) Novel yellow pigment from *Pterocarpus santalinus*: biogenetic hypothesis of Santalin analogues. Tetrahedron Lett 36:5599–5602

Kondeti VK, Badri KR, Maddirala DR, Thur SKM, Fatima SS, Kasetti RB et al (2010) Effect of *Pterocarpus santalinus* bark, on blood glucose, serum lipids, plasma insulin and hepatic carbohydrate metabolic enzymes in streptozotocin-induced diabetic rats. Food Chem Toxicol 48:1281–1287

Krishnaveni KS, Rao JVS (2000a) An isoflavone from *Pterocarpus santalinus*. Photochemistry 53:605–606

Krishnaveni KS, Rao JVS (2000b) A new acylated isoflavone glucoside from *Pterocarpus santalinus*. Biol Pharm Bull 23:1373–1374

Krishnaveni KS, Rao JV (2000c) A new triterpene from callus of *Pterocarpus santalinus*. Fitoterapia 71:10–13

Krishnaveni KS, Rao JV (2000d) A new isoflavone glucoside from *Pterocarpus santalinus*. J Asian Nat Prod Res 2(3):219–223

Kukla AS, Kumar N, Sanduja SK, Seshadri TR (1976) Sesquiterpenoids of *Pterocarpus santalinus* heartwood absolute configuration of pterocarpol. Indian J Chem Sect B Organc Chem Incl Med Chem 14:905–906

Kumar N, Seshadri TR (1975) Triterpenoids of *Pterocarpus santalinus*, constitution of a new lupine diol. Phytochemistry 14:521–523

Kumar N, Seshadri TR (1976a) Terpenoids of *Pterocarpus santalinus* leaves. Curr Sci 45 (14):516–517

Kumar N, Seshadri TR (1976b) A new triterpene from *Pterocarpus santalinus* bark. Phytochemistry 15:1417–1418

Kumar N, Ravindranath B, Seshadri TR (1974) Terpenoids of *Pterocarpus santalinus* heartwood. Phytochemistry 13:633–636. https://doi.org/10.1016/S0031-9422(00)91365-7

Kumar AS, Mohan EM, Gandhimathi R, Amudha P (2009) Study on the anti-seizure activity of methanolic extracts of *Indigofera tinctoria* L. Pharmacol Online 1:1341–1351

Kurzer MS (2000) Hormonal effects of soy isoflavones: studies in premenopausal and postmenopausal women. J Nutr 130:660S–661S

Langcake P, Pryce RJ (1977) A new class of phytoalexins from grapevines. Experientia 33 (2):151–152

Li L, Tao RH, Wu JM, Guo YP, Huang C, Liang HG, Fan LZ, Zhang HY, Sun RK, Shang L, Lu LN, Huang J, Wang JH (2017) Three new sesquiterpenes from *Pterocarpus santalinus*. J Asian Nat Prod Res 5:1–7

Liu T, Li Y, Sadiq FA, Yang H, Gu J, Yuan L, Lee YK, He G (2018) Predominant yeasts in Chinese traditional sourdough and their influence on aroma formation in Chinese steamed bread. Food Chem 242:404–411

Patrick D, Delphine A, Serge F (1998) Cosmetic compositions comprising santalins and/or santarubins for artificially colouring the skin. WHO 9844902 A1

Pueppke SG (1996) The genetic and biochemical basis for nodulation of legumes by rhizobia. Crit Rev Biotechnol 16:1–51

Raju KK, Nagaraju A (1999) Geobotany of Red Sanders (*Pterocarpus santalinus*): a case study from the southeastern portion of Andhra Pradesh. Environ Geol 37:340–344

Raju KK, Raju AN (2000) Biogeochemical investigation in south eastern Andhra Pradesh: the distribution of rare earths, thorium and uranium in plants and soils. Environ Geol 39 (10):1102–1106

Ramawat KG, Goyal S (2008) Indian herbal drugs scenario in global perspectives. Bioact Mol Med Plants 325–347

Ravindranath B, Seshadri TR (1972) Chemistry of santalin pigments 1. Structure of santalin permethyl ether. Tetrahedron Lett 13(13):1201–1204

Ravindranath B, Seshadri TR (1973) Structural studies on santalin permethyl ether. Phyto Chem 12 (11):2781–2788

Sawhney PL, Seshadri TR (1956) Special chemical components of commercial woods and related plant materials. IV. Phenolic components of *Pterocarpus* species. J Sci Ind Res 15C:79–82

Shankaranarayana KH, Prabhu VV, Theagarajan KS (1993) DARSE – a new essential oil from acidic alcohol extracts of red sanders. Indian Perfum 37:208

Singh S, Paliwal MK, Siddiqui IR, Singh J (1992) Two new coumarin glycosides from *Pterocarpus santalinus*. Fitoterapia 63:555

Singh S, Paliwla MK, Singh J (1993) A new prenylated coumarin from *Pterocarpus santalinus*. Fitoterapia 64:90

Tennekone K, Kumara GRRA, Kottegode IRM, Perera VPS, Weerasundara PSRS (1998) Sensitization of nanoporous films of TiO2 with santalin (red sandal wood pigment) and construction of dye-sensitized solid state photovoltaic cells. J Photochem Photobiol A Chem 117:137–142

Thimmappa R, Geisler K, Louveau T, O'Maille P, Osbourn A (2014) Triterpene biosynthesis in plants. Annu Rev Plant Biol 65:225–257

Van Rhijn R, Vanderleyden J (1995) The *Rhizobium*-plant symbiosis. Microbiol Rev 59:124–142

Venkataramaiah V, Prasad SV, Rajeswara RG, Swamy PM (1980) Level of phenolic acids in *Pterocarpus santalinus* L.f. Indian J Exp Biol 18:887–889

Vuky J, Motzer RJ (2000) Synthesis & pharmacological evaluation of coumarin derivatives as cannabinoid receptor antagonist & inverse agonist & reported 5-sub. 3-benzylcoumarin derivatives. Urol Oncol 5:240–249

Wu S-F, Chang F-R, Wang S-Y, Hwang S-Y, Hwang T-L, Lee C-L, Chen S-L, Wu C-C, Wu W-C (2011a) Anti-inflammatory and cytotoxic neoflavonoids and benzofurans from *Pterocarpus santalinus*. J Nat Prod 74:989–996

Wu S-F, Hwang T-L, Chen S-L, Wu C-C, Ohkoshi E, Lee K-H, Chang F-R, Wu Y-C (2011b) Bioactive components from heartwood of *Pterocarpus santalinus*. Bioorg Med Chem Lett 21:5630–5632

Yoganarasimhan SN (2000) Medicinal plants of India, Tamil Nadu, vol 2. Cybermedia, New Delhi

Yu P-L, Li-Hung Chen L-H, Chung K-R (2016) How the pathogenic fungus *Alternaria alternata* copes with stress via the response regulators SSK1 and SHO1. PLoS One. 10 11(2):e0149153. https://doi.org/10.1371/journal.pone.0149153

Colouring Principle (Dye) 5

T. Pullaiah and V. Damodara Reddy

Abstract

The fabulous tree *Pterocarpus santalinus* is known for its scarlet or purplish red colour wood, and this exquisite beauty of red sanders is due to the presence of colouring principles. Owing to its multitudinous applications, many researchers investigated the phytochemistry and applications in various sectors. The major pigment santalin is not only used as a dye but also has pharmacological properties. In this chapter authors provided a comprehensive review of the colouring principles of the red sanders and its multifarious uses.

Keywords

Santalin · Santarubin · Dye

5.1 Introduction

The natural dyes, especially vegetable colourants, have considerable interest in dyeing of textiles due to their perceived eco-friendly nature. During the last three decades, natural dyes have witnessed a process of revival due to increased awareness coupled with proven detrimental effects of synthetic dyes.

Pterocarpus santalinus is a small- to medium-sized deciduous tree. The bioprospecting efforts and available literature demonstrated the presence of several compounds in heartwood powder in particular santalins A, B and Y; pterocarpol;

T. Pullaiah (✉)
Department of Botany, Sri Krishnadevaraya University, Anantapur, Andhra Pradesh, India

V. Damodara Reddy
Department of Biochemistry, Sri Krishnadevaraya University, Anantapur, Andhra Pradesh, India

T. Pullaiah et al. (eds.), *Red Sanders: Silviculture and Conservation*,
https://doi.org/10.1007/978-981-13-7627-6_5

pterocarptriol; isopterocarpalone; pterocarpodiolone; cryptomeridol; and several nonspecific constituents such as isoflavones, isoflavonoid glucosides, triterpenes, sesquiterpenes, β-sitosterol, lupeol, epicatechin, lignans and pterostilbenes. In the present chapter, a systematic review about the colouring principles of red sanders is portrayed.

5.2 Colouring Principle

The advances in analytical techniques witnessed the isolation and characterization of various phytoconstituents present in red sanders. The group of extractable dyes, present in *P. santalinus*, are considered as water insoluble, as they are extracted either by using nonaqueous solvents or under alkaline extraction conditions (Surowiec et al. 2004). The major colouring principle of red sanders, santalin (santalic acid – a quinonoid moiety, constituting 16%), readily soluble in 90% alcohol but almost insoluble in water, was first isolated in its crude form in 1833. Santalin yields a blood red solution with alcohol, yellow with ether and violet with ammonia and caustic alkali (Puri and Seshadri 1955). This is partially soluble in some of the essential oils, such as lavender, rosemary, clove oil and oil of bitter almonds, and as a colouring agent, it forms part of the official composition of tincture of lavender (Krishnaveni and Rao 2000).

Santalin A, santalin B and deoxysantalin are major constituents of red- or purple-red- coloured natural dye extracted from the heartwood of Indian red sanders tree (*Pterocarpus santalinus*). This natural dye is used to colour food, alcoholic beverages, wood polish, metal varnishes, textiles, wool, silk, leather, jute, anti-sun tanning agents, hair dyes, medicines and tablet coating and as dye in sensitized solar cells.

The usage of powdered wood of red sanders for dyeing wool, cotton and leather and for staining other woods is featured in many publications including Wealth of India (1969). This revered textile dye from red sanders has five colouring components ranging from violet to orange, but the most popular is its bequeathing red colour (Gulrajani et al. 2002; Gulrajani and Bhaumik 2003; Siva 2003; Ferreira et al. 2004). In Europe the red dye is used as a colouring agent for medicinal preparations, while the French furniture makers during 1660–1885 used it for dyeing, polishing and varnishing (New 1981). In the USA, it is approved as a food dye and used for colouring alcoholic beverages and also approved within Europe where it is expended as a spice extract rather than food colourant (Mulliken and Crofton 2008).

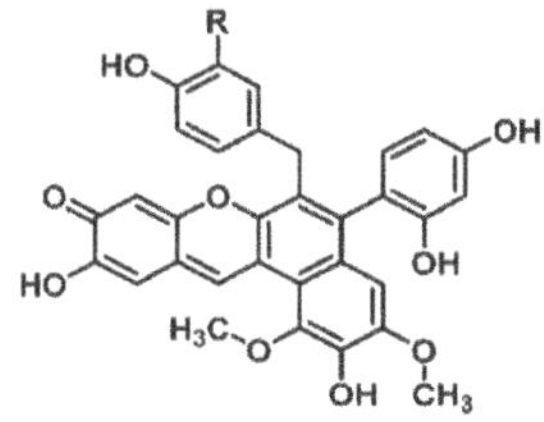

Chemical features of santalins A and B. Chemical structure of santalin (R = OH, santalin A; R = OCH_3, santalin B).

5.2.1 Santalin A

$C_{30}H_{17}O_7(OMe)_3$, formed red needles from aqueous methanol, melting point (m.p) 302–303 °C

λ_{max} (MeOH) 241, 280, 319, 471, 504 nm

γ^{KBr}_{max} 3636, 1639, 1613, 1538, 846, 784 cm^{-1}

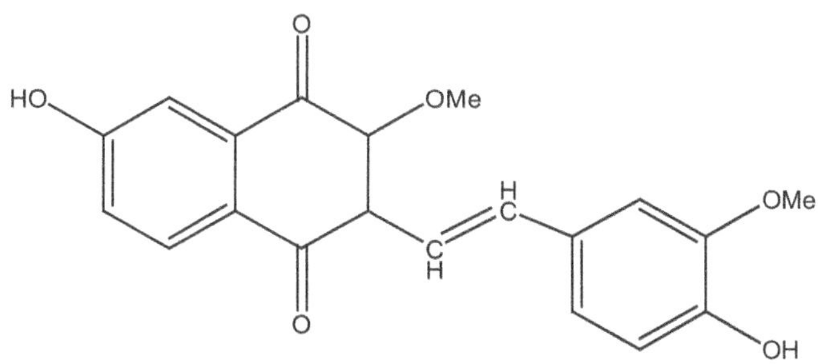

Deoxysantalin

Deoxysantalin

5.2.2 Santalin B

The red sanders (*Pterocarpus santalinus*) pigment santalin A has been identified as 6-(3,4-dihydroxybenzyl)-2,10-dihydroxy-5-(4-hydroxy-2-methoxyphenyl)-1,3-dimethoxybenzo[*a*]xanthen-9-one based on degradative and spectroscopic evidence. Santalin B is the corresponding 6-(4-hydroxy-3-methoxybenzyl) derivative (Arnone et al. 1975). NMR data, particularly those obtained by use of the shift reagent Eu $(dpm)_3$, have led to the assignment of a revised structure to per-*O*-methylsantarubin.

Santalin is tasteless, nearly insoluble in water, but soluble in alcohol, ether, acetic acid and alkaline solutions and slightly by the oils of lavender and rosemary. The alcoholic solution of santalin gives different coloured precipitates with metallic solutions; thus with tin it gives a beautiful purple, lead a violet, iron a deep brown and silver a reddish brown colour.

5.3 Extraction of the Dye

A natural dye is a substance derived from natural sources used to add a colour to or change the colour of something and considered as sustainable and eco-friendly. *Pterocarpus* species is renowned for its characteristic timber of gorgeous colour, magnificence and exceptional technical qualities yielding a natural dye santalin belonging to the molecular class of condensed bioflavonoid. Extraction of bioactive compounds is influenced by various process parameters such as solvent composition, pH, temperature, extraction time and solid to liquid ratio.

The deep red heartwood is utilized for extraction of the colouring principle. The process involves reduction of the wood to chips or powder and extraction with alcohol. The extract may be simply concentrated or be stripped of solvent to give a solid product prior to sale. No reliable published information is available on commercial extraction process. For sale to end-users, specific formulations as liquids, dispersed solids or water-soluble forms are prepared at strengths appropriate for the usage in different preparations.

The pigment of red sanders and of related red woods has attracted the interest of natural product chemists for more than a century since the first report by Pelletier (1833). Despite much effort, it was not until 1954 that two pure compounds santalin and santarubin were isolated by Robertsen and Whalley (1954). The presence of two pigments, santalins A and B, were reported by Ravindranath and Seshadri (1972). Santalins A, B and C were reported by Arnone et al. (1975). Arnone et al. (1975) isolated the santalin by extracting the heartwood shavings with hexane and then with chloroform and ethyl acetate. The ethyl acetate extract was concentrated, and the residue chromatographed through a woollen polyamide column ($CHCl_3$-MeOH) to yield red product of which santalins A and B were the major constituents. Arnone et al. (1975) separated mixture of santalins A and B along with one major and minor constituent by extracting the sapwood shavings with light petroleum, chloroform and ethanol in succession. Chloroform and ethanol extracts were combined, and the solvent is removed. The residue is dissolved in hot benzene and left for overnight in the refrigerator. After filtration the mother liquor was enriched with santalin and chromatographed over polyamide column to obtain santalins. Robertson and Whalley (1954) isolated santalins by extracting the powdered wood with carbon tetrachloride followed by diethyl ether. After evaporating the solvent ether, the extract is dissolved in ethyl acetate and dried to attain solid state. This solid contains santalins crystallization from alcohol yielded santalins.

Jain et al. (2009) patented an improved process for the extraction of total santalins from the heartwood of *Pterocarpus santalinus* without resorting to chromatography procedures. The main objective of this invention was to provide a simple economical process by which total santalins can be separated in powder form. Yet another objective is to provide a process in which no organic solvents are employed to separate and purify the total santalins from red sanders. Their invention provides an improved process for the preparation of total santalins from the heartwood of *Pterocarpus santalinus* (red sanders). The described process is pulverization of the heartwood and extraction in an alkaline medium at a suitable temperature ranging from 10 to 100 °C. The alkaline medium may be sodium hydroxide, potassium hydroxide, sodium carbonate, sodium bicarbonate, liquid ammonia, or in any alkaline medium and their mixtures. Thus, obtained extract is filtered, washed with water and acidified with organic acids, mineral acids or any weak acids like CH_3COOH, oxalic acid and tartaric acid in the pH range of 4.0–6.9. This acidified mixture is kept for precipitation in a refrigerator; the precipitate is then filtered, washed with water and dried under pressure to get powdered form of santalins.

Gulrajani and Bhaumik (2003) reported the red sander extraction process and its use for dying wool and nylon. They also mentioned the possibility of extracting the colouring matter from the wood with organic as well as aqueous alkali medium with the same absorption maxima up to pH10 in the methanol. The dye extracts can be used either directly on wool and nylon or along with different mordants, and the colour tone changes with different metal ions in both wool and nylon. The lightfastness of wool and nylon samples dyed with red sandalwood extract is poor than that of those dyed with synthetic dyes. However it is good when mordants like copper sulphate, ferrous sulphate and potassium dichromate (post-mordanting) are used but poor with alum and alum + tartaric acid. In the case of nylon, the lightfastness is very poor. Wash fastness of all the samples dyed and treated with mordants except copper sulphate is good for both wool and nylon.

In recent years various extraction methods are used for efficient extraction of phytoconstituents. Among the most common practice is the usage of organic solvents such as hexane, ethyl acetate, ether, chloroform, benzene, ethanol and methanol. Solvent extraction method is typically used in combination with other technologies such as solidification/stabilization, precipitation and electrowinning. Another distinctive method called ultrasound-assisted extraction (UAE) is adopted, and this is a process of high extraction, which yields good quality of phytocompound in shorter periods of time consuming lesser amounts of solvent than traditional processes. Among the new extraction techniques, UAE is the most economical and the one with less instrumentation requirements. Different plant extracts and bioactive metabolites have been obtained with this technique. Microwave-assisted extraction (MAE) is a relatively latest extraction technique which utilizes microwave energy to heat the solvent and the sample to increase the mass transfer rate of the solutes from the sample matrix into the solvent. Microwave extraction offers better selectivity, less extractant use and lower energy input. MAE offers a rapid delivery of energy to a total volume of solvent and solid target matrix with subsequent heating of the solvent and solid matrix, efficiently and homogeneously.

Hemanthraj et al. (2014) studied optimization of extraction parameters for natural dye from *Pterocarpus santalinus* by using response surface methodology. In this study, response surface methodology was used to optimize the solvent-, ultrasound- and microwave-assisted extraction of natural dye from *Pterocarpus santalinus* wood. Box-Behnken design was used to determine the optimum process parameters and the multiple regression analysis for predicting responses. Under optimum conditions microwave-assisted extraction method showed the highest natural dye yield percentage which is 50.0 for ethyl acetate and 50.2 for methanol. Microwave-assisted extraction method dictates the quality, economics and environmental impact of any processing plant. It shows a highly promising future with drastic reduction in extraction time resulting in higher sample throughput without significant losses in analyte recovery. In this study among three parameters, solid to liquid ratio for solvent extraction method, extraction time for ultrasonic extraction method and extraction time and solid to liquid for microwave-assisted extraction method is found to be most prominent factor affecting the efficiency of dye extraction.

Almost 200 years after Pelletier's pioneering studies on the chemical constituents of red sanders, the major santalins and santarubins have been synthesized. This efficient approach integrates a Knochel isoflavonoid synthesis with Friedel-Crafts alkylations or olefin metatheses, and a final biomimetic reaction cascade that furnishes the venerable benzoxanthenone dyes in a single operation (Strych and Trauner 2013).

5.4 Applications

"Red sandalwood" or "red sanders" is a member of the "insoluble redwoods", i.e. red dye stuff sources which are not extractable by water (*Pterocarpus indicus* of South and Southeast Asia and *Pterocarpus* species of Africa). Red sandalwood and some of the other "insoluble redwoods" were employed in the past for dyeing wool, cotton and leather and for wood staining. As mentioned earlier, the principal red pigments in red sanders heartwood are santalins A and B, soluble in organic solvents and alkalis but not in water. Today, usage of red sanders appears restricted to foods where it imparts a sweet-spicy flavour and orange-red shades. For the food industries, the extract is normally sold in the alcohol-soluble form, either as liquids or powders; however water-soluble forms (salts) are also made available. Red sanders has been traditionally used in preparation of fish products in Europe, and more recent applications include the colouring of seafood sauces, meat products, snack foods, breadcrumbs and alcoholic drinks. Dosage levels range from 0.1% to 1% (weight for weight). A yellow isoflavone pigment, santal, is also present. The santalins together with other related pigments are found in some other species of *Pterocarpus* and *Baphia* (Green 1995). Red sandalwood has traditionally been used as an astringent, disinfectant or diuretic, but recently it is used as a natural dye to colour herbal mixtures and toothpaste and has become one of the most popular natural dyes used today.

P. santalinus is highly valued for its heavy, dark claret-red heartwood, which yields 16% of red colouring matter santalin, and is used as colouring agent in pharmaceutical preparations and food stuff (Arunakumara et al. 2011). Red sanders is used in cosmeceutical preparations to impart glow and shine to the skin. These are principal ingredients of creams, gels, lipsticks and lotions and found to have wound healing properties. The dye principles not only have cosmetic value but also are employed for nutraceutical applications and hence added while processing food beverages, candies, wine and other alcoholic beverages. A histological stain prepared from the heartwood of *P. santalinus* has been found to be an excellent nuclear stain for various cells of animals and plants. As an elastic tissue stain, the results are comparable to standard elastic tissue stains. The stain can be used as counter stain with certain histochemical procedures with satisfactory results (Sen Gupta and Mukherjee 1981).

5.5 Suppliers of Red Sanders Dye

Source: http://www.tradeboss.com

1. Sri Balaji Biotec Industries, 20-1-277, Korlagunta, Tirupati-517502, Andhra Pradesh, India, Telephone: 91-866-2441-279; Fax: 91-866-2441-279 http://www.sribalajibiotec.com/
2. Whole Herb Company, 19800 8th St. East Sonoma, California-95476, USA. Tel:707-935-1077 Fax: 707-935-3447; www.wholeherbcompany.com www.berjeinc.com
3. Natural Red 22 | BESTCHEM Hungária Kft
4. BESTCHEM Hungaria Kft, https://bestchem.hu/bestchem/en/chemicals/dyes-pigments/item/NR22
5. Sriya Associates, #54-11/43/3 Aditya Nagar, Krishna College Road, HB Colony, Visakhapatnam-530017, Andhra Pradesh, India. Ph: 091-9347924239; Website: http://www.redsandersexport.webs.com; e-mail: sriyaexim@gmail.com
6. Hangzhou Henghao Pigment Chemical Co., Ltd. Zhejiang, China.

References

Arnone A, Carnara L, Merlini L, Nasini G (1975) Structures of Red sandalwood pigments santalins A and B. J Chem Soc Perkin Trans 1(2):186–194

Arunakumara KKIU, Subasinghe BCWS, Yoon MH (2011) *Pterocarpus santalinus* L.f. (Roth handgun): a review of its botany, uses, phytochemistry and pharmacology. J Korean Soc Appl Biol Chem 54:495–500

Ferreira ESB, Alison N, Hulme AN, McNab H, Quye A (2004) The natural constituents of historical textile dyes. Chem Soc Rev 33:329–336

Green CL (1995) Natural colorants and dye stuff. A review of production, market and developmental potential. FAO, Rome, p 16

Gulrajani ML, Bhaumik S (2003) Dyeing of red sandal wood on wool and nylon. Indian J Fibre Text Res 27:221–226

Gulrajani ML, Bhaumik S, Oppermann W, Hardtmann G (2002) Kinetic and thermodynamic studies on Red Sandalwood. Indian J Fibre Text Res 27:91–94

Hemanthraj KPM, Desai SM, Bisht SS (2014) Optimization of extraction parameters for natural dye from *Pterocarpus santalinus* by using response surface methodology. Intern J Eng Res Appl 4(9):100–108

Jain SM, Qazi GN, Suri OP, Gupta DK (2009) An improved process for the extraction of total santalins from the heartwood of plants *Pterocarpus santalinus* L. India Patent, Patent number 227103
Krishnaveni KS, Rao JVS (2000) A new acylated isoflavone glucoside from *Pterocarpus santalinus* Chem Pharm Bull 48(9):1373–1374
Mulliken T, Crofton P (2008) Review of the status, harvest, trade and management of seven Asian CITES-listed medicinal and aromatic plant species. BFN, Skripten, Bonn, p 227
New S (1981) The use of stain by furniture makers: 1660 to 1885. Furnit Hist 17:51–60
Pelletier J (1833) Untersuchung über die elementare Zusammensetzung mehrerer näheren Bestandtheile der Vegetabilien. Annalen 6:21–34
Puri B, Seshadri TR (1955) Survey of anthoxanthins. Part IX. Isolation and constitution of palasitrin. J Sci Ind Res 14B:1589–1592
Ravindranath B, Seshadri TR (1972) Chemistry of the santalin pigments part 1 structure of santalin per methyl ether. Tetrahedron Lett 13:1201–1204
Robertsen A, Whalley WB (1954) The chemistry of the "insoluble red" woods. Part VI. Santalin and santarubin. J Chem Soc 76:2794–2801
Sen Gupta PC, Mukherjee AK (1981) Newer applications of the histological stain prepared from *Pterocarpus santalinus*. Stain Technol 56(2):79–82
Siva R (2003) Status of natural dyes and dye-yielding plants in India. Curr Sci 92:916–925
Strych S, Trauner D (2013) Biomimetic synthesis of santalin A and B and santarubin A, B, the major colorants of Red sandalwood. Angewandtee Chem. https://doi.org/10.1002/anie.201302317
Surowiec I, Nowik W, Trojanowicz M (2004) Identification of insoluble red dyewoods by high performance liquid chromatography photodiode array detection (HPLC-PDA) fingerprinting. J Sep Sci 27(3):209–216
Wealth of India (1969) A dictionary of Indian raw materials and industrial products. CSIR, New Delhi

Pharmacology of Red Sanders

6

T. Pullaiah and V. Damodara Reddy

Abstract

A wide array of biological activities and potential health benefits of *P. santalinus* have been reported including antioxidative, antidiabetic, antimicrobial, anticancer, anti-inflammatory properties as well as protective effects on the liver, gastric mucosa and nervous systems. All these protective effects were attributed to bioactive compounds present in *P. santalinus*. The present chapter describes the pharmacological effects of *P. santalinus* in health and disease with an update.

Keywords

Cosmeceuticals · Antioxidant · Antidiabetic · Antimicrobial · Anticancer · Anti-inflammatory · Hepatoprotective · Wound healing

6.1 Introduction

Recently there has been increasing interest towards utilization of plant and plant-derived compounds in nutraceutical and pharmaceutical preparations. Ayurveda, an Indian traditional medicine, described wide spectrum of medicinal properties of *Pterocarpus santalinus* (*P. santalinus*). Many important bioactive phytocompounds were extracted and identified from heartwood of *P. santalinus*. Bioactive compounds typically occur in small amounts, and they have more subtle effects than nutrients. These bioactive compounds influence cellular activities that modify the risk of disease and cure and alleviate disease symptoms. The bioactive compounds have potentially important health benefits, and these compounds can act as antioxidants,

T. Pullaiah (✉)
Department of Botany, Sri Krishnadevaraya University, Anantapur, Andhra Pradesh, India

V. Damodara Reddy
Department of Biochemistry, Sri Krishnadevaraya University, Anantapur, Andhra Pradesh, India

T. Pullaiah et al. (eds.), *Red Sanders: Silviculture and Conservation*,
https://doi.org/10.1007/978-981-13-7627-6_6

enzyme inhibitors and inducers, inhibitors of receptor activities and inducers and inhibitors of gene expression, among other actions. A wide array of biological activities and potential health benefits of *P. santalinus* have been reported including antioxidative, antidiabetic, antimicrobial, anticancer, anti-inflammatory properties as well as protective effects on the liver, gastric mucosa and nervous systems. Arunakumara et al. (2011) reviewed the phytochemistry and pharmacology of *Pterocarpus santalinus*. The present chapter describes the pharmacological effects of *P. santalinus* in health and disease with an update.

6.2 Medicinal Properties and Uses

The heartwood of *P. santalinus* is considered as cooling, astringent, antipyretic, diaphoretic, febrifuge and tonic in action. It is employed in drug formulations used in the treatment of dysentery, diabetes, stomach ulcers, bilious affections and diseases of blood and as an anthelmintic and diaphoretic. The wood paste is externally applied to cure skin inflammation, headache, fever, scorpion sting and skin diseases and to improve eyesight. It has remarkable property of healing pimples, scars, boils, wounds, burnt marks, black spots, eczema and other blemishes of the skin (Pullaiah 2006; Parrotta 2001; Padmalatha and Prasad 2006). A red chemical substance, Santalin, present in the heartwood of this plant is used as a dyeing/staining agent. In European medicine, it is used as a colouring agent. The fruit decoction is used as astringent tonic in chronic dysentery (Kondeti et al. 2010). The stem and bark extracts are anthelmintic, aphrodisiac and alexiteric. The stem bark and leaf extracts have antibacterial properties and are effective against bacterial infection (Azamthulla et al. 2015). The paste of the wood is considered as coolant and applied externally for treating inflammations and healing wounds (Krishnaveni and Srinivasa Rao 2000). The wood extracts are used in treatment of cough, vomiting, thirst, eye diseases, ulcers and diseases of the blood (Kondeti et al. 2010). The wood paste is considered for curing bone fracture, leprosy, spider poisoning, scorpion sting, blood vomiting and mental aberrations. The extracts are used for blood purification, alleviating digestive problems and treating blood pressure and bleeding piles and to control haemorrhage (Azamthulla et al. 2015; Bhagyaraj 2017; Ramabrahmam and Sujatha 2016).

6.3 Cosmeceuticals

Red sandalwood powder which also known as Rakta Chandana powder has been used in Ayurvedic medicine as an antiseptic, wound healing and anti-acne agent. Red sandalwood powder can be made into a paste with honey and water and can be applied to the face. It can also be combined with ashwagandha and cosmetic clays or mud for a detoxifying facial. Red sandalwood powder is used in soap to yield a dark maroon to purple colour.

Red sandalwood or 'Rakta Chandana' is known to be one of the most reliable natural beauty ingredients. When used as a face pack or face mask, it can benefit the skin in multitudinous ways. It helps in fading the marks caused due to acne and improves complexion and facial glow. It is an effective anti-tanning agent, even alleviates acne and pimples and is a very good skin toner.

The wood powder is mixed with tomato juice and used as a face pack to remove tan and lighten skin naturally. This also removes acne and dark spots effectively with regular use. After mixing with milk and honey, the paste can be employed to manage dry skin, by spreading over the face and neck and washing with cold water after 20 min.

Red sandalwood powder is mixed with rose water to make a face pack. Mixture of red sandalwood powder and rice powder can be used as a facial exfoliating scrub. Applying the paste over the skin and gently scrubbing in circular motion with fingertips clear blackheads. Red sandalwood powder mixed with cabbage juice, when applied as a face pack, removes scars and tightens wrinkles (Reddy 2018).

Red sandalwood paste along with turmeric and water can be used to cure pimples. Addition of camphor to this formulation can bring about a cooling effect. Reddy (2018) claims that by applying this paste to the pimples, it can help alleviate within a few hours. A combination of coconut oil, almond oil and red sanders powder when applied onto the face helps to tone the skin and also remove the sun tan (Reddy 2018).

Red sanders is also used for massage, reflexology and cosmetics, and its cooling effect calms the skin. Red sanders oil soothes and helps relieve from uncomfortable irritated symptoms. Oil of red sanders can be prepared by adding the powder to any oil of interest and need to be set for night. Though oil can be extracted by pressing the inner bark of the red sandalwood tree using water or alcohol, most of the extracted products available in the market are in powder form (Reddy 2018).

It is an age-old effective remedy to fight acne, skin rashes, sunburn, blemishes and premature aging and can be used in skin and healthcare formulations. Applying red sanders wood powder and green tea mask will reduce wrinkles and sun-induced aging. This can be prepared at home by mixing 1 tablespoon of powder and enough green tea water to make a paste. This paste when applied tightens the pores, and the natural antioxidants help in improving the signs of ageing.

Reddy (2018) gave various tips of using red sanders for cosmetic purposes, and they are as given under:

1. Add few drops of coconut oil to red sandalwood powder and make a paste with it. This can be applied on dry areas as moisturizer. Wash off after 10–15 min. Red sandalwood powder is known to provide nourishment to the skin cells.
2. Mix red sandalwood powder with lemon juice to prepare a mask for oily skin. Apply this paste all over your face and let it dry out. Once it does, wash it off with

lukewarm water. It helps in regulating the secretion of sebum and makes the pores tight.

3. Acne and pimples happen to be common skin problems faced by most people. A rose water and red sandalwood face pack helps in the reduction of acne and acne scars and the irritation caused by acne due to its cooling properties. You can even add a teaspoon of honey and a pinch of turmeric to the pack for better results.
4. A pack made with 1 tablespoon of red sandalwood and 2 tablespoons of mashed ripe papaya can help in exfoliation (removal of dead skin). This face pack helps you shed the dead skin and leaves your skin feeling fresh and rejuvenated.
5. Use red sandalwood powder along with curd and milk to improve uneven skin tone. Mix 1 tablespoon of red sandalwood powder, half tablespoon of turmeric and 2 tablespoons of curd and milk each to make a face pack. Apply this all over and let it dry. Later wash it off with lukewarm water and enjoy a radiant complexion.
6. It helps in removal of dark spots and reduces pigmentation. Prepare a simple pack with 2 tablespoons of red sandalwood and 2 tablespoons of milk, and apply it every day.
7. A pack with cucumber juice or curd with red sandalwood powder helps in removing sun tan. Mix 2 tablespoons of curd or cucumber juice with an equal amount of red sandalwood powder, and apply it on the affected area. Let it dry and then wash off. You will see instant results.
8. Prepare a mask with 2 teaspoons of almond oil, 4 teaspoons of coconut oil and 4 teaspoons of red sandalwood powder. Use this regularly for soft and glowing skin.

Red sandalwood powder is used as basic flavouring agent to yield a dark maroon to purple colour soap, and the recommended quantity per pound of soap is a teaspoon of red sandalwood powder (Bhagyaraj 2017).

Red sandalwood diluted with water can be used as a body spray which will balance body's heat and also reduces skin transmissions (Ramabrahmam and Sujatha 2016). Red sandalwood combined with milk and honey applied as paste and rinsed after 20 min with cold mineral water is effective for dry skin (Ramabrahmam and Sujatha 2016). Red sandalwood powder is made into paste using cosmetic clays or mud, ashwagandha and honey with water for detoxifying the face (Bhagyaraj 2017).

Paste of red sandalwood powder with turmeric and water/rose water is applied to pimples during night as a treatment. This paste, if camphor is added, alleviates pimples within a few days (Ramabrahmam and Sujatha 2016; Bhagyaraj 2017). This also helps to reduce irritation of ripe pimples and skin rashes (Ramabrahmam and Sujatha 2016).

Mixture of coconut oil and almond oil along with sandalwood powder removes sun tan and balances the skin tone (Bhagyaraj 2017). Combination with tomato juice, when used as a face pack, reduces tan and helps clear and lighten skin tone (Ramabrahmam and Sujatha 2016) (Figs. 6.1 and 6.2).

Fig. 6.1 Face pack with red sandalwood. (Courtesy Reddy 2018 with permission)

Fig. 6.2 Skin rejuvenating red sanders cream. (Courtesy Reddy 2018 with permission)

6.4 Anti-melanogenic Activity

Tyrosinase is a copper-containing enzyme and is widely distributed in animals, plants and microorganisms. The enzymes showed considerable structural similarity independent to the kingdom they belong. Tyrosinase is a key enzyme in melanin biosynthesis, involved in determining the colour of the mammalian skin and hair. Increased activity of the enzyme can lead to hyperpigmentation resulting in distressing aesthetic values. The inadequacy of current conventional methods to inhibit tyrosinase activity safely encourages the need to seek new potent tyrosinase inhibitors in cosmetic and therapeutic applications. Prasad et al. (2016) reported the effectiveness of hot water extract of *Pterocarpus santalinus* bark against the melanin-producing system of *Bacillus cereus*. The extract had shown to inhibit melanin production in bacteria dose dependently. Their results suggested that *P. santalinus* extract possesses anti-melanogenic/antityrosinase activity, which could be utilized as a safe depigmentation agent (Prasad et al. 2016).

6.5 Radical Scavenging and Antioxidant Activity

Free radicals are electrically charged molecules, an excess generation of free radicals linked to many human diseases. Reactive oxygen species (ROS) and hydroxyl radicals cause oxidation of lipids, proteins and DNA, damage the structure and function of cells and lead to development of diseases. Animal studies have shown that dietary phytochemical antioxidants can remove free radicals. Among them, phenolic and polyphenolic compounds, such as flavonoids in edible plants, exhibit potent antioxidant activities (Huang et al. 2009). The free radical scavenging activity of extracts of leaves of *P. santalinus* has been evaluated using in vitro studies. Methanolic extract of leaves exhibited 1,1-diphenyl-2-picrylhydrazyl (DPPH), nitric oxide and hydrogen peroxide radical scavenging activity (Arokiyaraj et al. 2008; Krishnamoorthy et al. 2011). Studies of Kumar (2011) demonstrated that methanolic extract of heartwood showed ferric ion (Fe^{3+})-reducing capacity and DPPH radical scavenging activity. The reductive capabilities and scavenging activity were found to increase with the increasing concentrations and were compared with butylated hydroxyanisole (BHA) (Kumar 2011). Reports also revealed that pterostilbene exhibited strong in vitro antioxidant activity against free radicals such as DPPH, 2,2′-azino-bis(3-ethylbenzothiazoline-6-sulfonate) (ABTS), hydroxyl, superoxide and hydrogen peroxide (Achary and Ghaskhadbi 2013). Very limited reports were available related to scavenging activities of extracts isolated by using different solvents and isolated active compounds, which will be needed further elaborate and confirmative studies using heartwood of *P. santalinus*.

6.6 Antihaemolytic Activity

Lokesh et al. (2016) investigated the qualitative and quantitative aspects of crude phytocompounds of *Pterocarpus santalinus* heartwood and its effect on ethanol-induced haemolysis in human subjects. In vitro studies on different subjects showed that aqueous extracts possess antioxidant and haemolytic activity in the following order - Smokers > alcoholic smokers > alcoholics. Their data showed that the active compounds may offer protection against free radical-mediated oxidative stress in RBC of humans with ethanol-induced haemolysis.

6.7 Antibacterial Activity

Infectious diseases are major reason for morbidity and mortality globally. It accounts roughly 50% of all deaths in tropical countries (Lopez et al. 2006). Chemical and synthetic drugs used to treat microbial infections are associated with multiple side effects. After routine usage human pathogens have developed drug resistance, and it is a big challenge for researchers to develop safe and effective medication for the treatment of infectious diseases. Several research groups studied the antibacterial and radical scavenging activities of plant extracts and suggested that extracts showed effective antibacterial and radical scavenging activities with no side effects (Faremi et al. 2008; Donald. 2000). The antibacterial activity of leaf and bark extract of *P. santalinus* was tested against Gram-positive bacteria such as *Staphylococcus* [ATCC-25933], *Staphylococcus epidermidis* [MTCC-3615], *Bacillus subtilis* [MTCC-441], *Enterococcus faecalis* [ATCC-2912], *E.coli* [ATCC-25922] and *Pseudomonas aeruginosa* [ATCC-27853]. The broad-spectrum antibacterial activity exhibited by *P. santalinus* may be attributed to its richness in isoflavone glucosides (Krishnaveni and Srinivasa Rao 2000). Flavonoids are known to be synthesized by plants in response to microbial infections, and therefore, very obviously, they have been found in vitro to be effective antimicrobial substance against a wide range of microorganisms. Catechins, an important group of flavonoids, have been extensively investigated due to their occurrence in oolong green tea. It has been reported in the past that tea possess antimicrobial activity and that they contain a mixture of catechin compounds (Ullah and Khan 2008). Bark extract of *P. santalinus* showed the maximum inhibition activity than leaf extract (Krishnamoorthy et al. 2011; Manjunatha. 2006a). Antibacterial activity of 21 timber-yielding plants was tested against uropathogenic bacteria, and *P. santalinus* is one of the important plants among maximum inhibitory activity-exhibiting plants (Mishra and Padhy 2013). A comparative study was done to know the antibacterial activity of methanol and aqueous leaf extract of *P. santalinus.* Antimicrobial activity of leaf extract from *P. santalinus* exhibited significant antimicrobial activity at all the dosage tested (1.25 mg/disc, 2.5 mg/disc and 5 mg/disc). Leaf extract of *P. santalinus* can be a potential source of new antimicrobial agents for the tested drug-resistant bacterial strains and fungi (Balaraju et al. 2008). Methanolic extract showed high inhibitory activity than aqueous extract (Dey et al. 2014). Silver nanoparticles prepared from leaf extract of *P. santalinus* exhibited antibacterial potential against Gram-positive and Gram-negative strains (Gopinath et al. 2013). Other reports also have shown that lignans present in *P. santalinus* exhibit antibacterial properties (Yamauchi et al. 2005). Anti-*Helicobacter pylori* effect was tested by culturing *H. pylori* on gastric epithelial cells in the presence/absence of *P. santalinus* extract. A reduction in the activity of urease, normal appearance of gastric epithelial cells under electron microscopic examination, decrease in LPO and lactate dehydrogenase (LDH) activity suggested anti-pyloric effect (Narayan et al. 2007a). In vitro antimicrobial activities of heartwood extracts needed standardization and confirmation against various pathological microbes.

Vandita et al. (2013) reported the effect of phytochemical constituents of *Pterocarpus santalinus* on antibacterial and antifungal activities.

Antimicrobial activity of leaf, stem and bark decoction extracts of *P. santalinus* against four Gram-positive bacteria and four Gram-negative bacteria was investigated by Donga et al. (2017a). The stem and bark extracts of *P. santalinus* inhibited all the four Gram-positive bacteria; however the leaf extract is not effective in inhibiting any bacteria. All the tested four bacterial strains are resistant to leaf extract. The stem extract showed slightly higher zone of inhibition as compared to the bark extract. The leaf, stem and bark extracts of *P. santalinus* inhibited *Pseudomonas aeruginosa*, while *Escherichia coli* and *Klebsiella pneumoniae* were inhibited by leaf and bark extracts. *Salmonella typhimurium* was inhibited by stem and bark extracts. Among the three parts, bark extract showed best antibacterial activity, while leaf extract showed minimum activity. Among eight Gram-positive and Gram-negative organisms, *Pseudomonas aeruginosa* was the most susceptible organism. Among the three parts, the best antimicrobial activity was shown by bark extract. All the three parts showed synergistic antimicrobial activity with antibiotics, but their levels varied. The results suggest that all the three parts enhance the antimicrobial efficacy of the antibiotics against some microorganisms and hence can be developed as a new therapeutic weapon against infectious diseases.

6.8 Antifungal Activity

Mohandass et al. (2010) and Arokiyaraj and Perinbam (2010) studied the antifungal activity of *Pterocarpus santalinus*. Leaves were exhaustively extracted with different solvents like hexane, ethyl acetate and methanol in ascending order of polarity. All the three extracts were subjected to antifungal screening. The ethyl acetate extract showed significant MIC values against *Trichophyton rubrum* (62.5 μg/ml), *Trichophyton mentagrophytes* (500 μg/ml), *Trichophyton simmi* (125 μg/ml), *Epidermophyton floccosum* (500 μg/ml) and *Scropulariopsis* sp. (500 μg/ml) (Mohandass et al. 2010)

Donga et al. (2017a) studied the antifungal activity of leaf, stem and bark extracts. The leaf, stem and bark extracts of *P. santalinus* inhibited *Candida albicans*, while *Candida glabrata* was inhibited by leaf and stem extracts. *Cryptococcus neoformans* and *Candida epicola* were inhibited by only stem extract. *Candida albicans* was the most susceptible fungal strain. Stem and bark extracts showed good antifungal activity.

6.9 Anti-inflammatory Effect

Inflammation plays a crucial role in the initiation and progression of pathological conditions. A majority of phytomedicines exhibit therapeutic efficacy by exerting anti-inflammatory and cytotoxic activities. It is well known that T-lymphocytes enhance chronic inflammatory condition by activating the inflammatory cells such as mast cells, eosinophils, neutrophils and macrophages resulting in massive production of chemical mediators and pro-inflammatory cytokines (Cho et al. 2001; Panayi et al. 1992). The anti-inflammatory activities of lignan compound isolated from heartwood of plant have been studied in RAW264.7 cells and splenocytes. Tumour necrosis factor (TNF)-α is one of the pro-inflammatory cytokines secreted during chronic inflammatory and allergic diseases. TNF-α production in lipopolysaccharide-stimulated RAW264.7 cells was inhibited at a dose of 25 μg/ml by the savinin, a lignan compound. Molecular mechanism of inhibitory action of the compound is due to the structural similarity with that of butyrolactone ring and its polarity on the c-9 position (Cho et al. 2001). Specific lignans, viz. savinin, calocedrin and eudesmin, were reported from heartwood extract of *P. santalinus*. These compounds were found to inhibit TNF-α and showed anti-proliferative effect with IC_{50} value at 40 μg/ml (Kwon et al. 2006). Studies showed a significant inhibitory activity of heartwood extract (500 μg/ml) against carrageenan induced anti-inflammation in paw oedema (Achary and Ghaskadbi 2013). Five new benzofurans and six neoflavonoids and pterolinus were isolated from heartwood; among them pterolinus B showed potent anti-inflammatory activity with IC_{50} value at 0.19 μg/ml (Wu et al. 2011a). Phenanthrenedione (pterolinus-K) and chalcone (pterolinus-L) compounds isolated from heartwood of *P. santalinus* showed notable inhibition of superoxide anion generation and elastase release by human neutrophils in response to fMLP/CB with IC_{50} values at 0.99 and 0.94 μM, respectively (Wu et al. 2011b). However, other biological activities and anti-inflammatory mechanisms exhibited by lignans and savinins need further studies. Bark wood extract powder showed significant anti-inflammatory and mild analgesic activity on topical application in rat model of chronic inflammation (Dhande et al. 2017).

6.10 Effect on Diabetes

The molecular mechanisms involved in the treatment of diabetes by individual as well as whole extract of either bark or heartwood of *P. santalinus* need to be thoroughly investigated. Traditionally made cups with heartwood have been used for drinking water as a treatment of diabetes (Panayi et al. 1992; Nagaraju and Rao 1989). Diabetes mellitus (DM) is a metabolic syndrome that consists of ineffective insulin regulation leading to derangements in carbohydrate metabolism (Nagaraju et al. 1991). The bark extracts of *Pterocarpus santalinus* at a dose of 0.25 g/kg body weight/day showed maximum antihyperglycaemic activity but failed to produce any hypoglycaemic activity in normal rats (Rao et al. 2001). The antihyperglycaemic activity of ethanolic extract of *P. santalinus* bark at the dose of 0.25 g/kg body weight was found to be more effective than that of glibenclamide (Rao et al. 2001). Active fraction of ethyl acetate/methanol (9:1) of bark ethanolic extract has showed anti-hyperglycaemic activity at a dose of 150 mg/kg b.wt. by improving insulin secretion and alterations in carbohydrate metabolism (Kondeti et al. 2010; Rao et al. 2001). Co-administration (250 mg/kg b.wt) of heartwood aqueous extract along with vitamin E to STZ-induced diabetic rats caused significant lowering of blood sugar at a dose of 250 mg/kg b.wt/day (Halim and Misra 2011). Phytochemical analysis of ethanolic extract of bark of *P. santalinus* active fraction showed the presence of flavonoids, glycosides and phenols. Bark extract effectively decreases streptozotocin-induced hyperglycaemia by increasing glycolysis and decreasing gluconeogenesis (Kondeti et al. 2010). Several researchers paid much attention to study effect of *P. santalinus* against diabetes using classical methods. However, future studies have been required to explore the antidiabetic activity using isolated bioactive compounds at molecular level.

6.11 Anticancer Activity

Natural phenolic compounds act on multiple targets in pathways and mechanisms related to carcinogenesis, tumour cell proliferation and death, inflammation, metastatic spread, angiogenesis and radiation resistance. Chemoprevention with food phytochemicals is meanwhile regarded as one of the most important strategies for cancer control. Phenolic compounds from medicinal herbs include phenolic acids, flavonoids, stilbenes, lignans and others (Huang et al. 2009). Bioactive compounds isolated from *P. santalinus* exhibit cytotoxic and anticancer properties. Treatment of cervical adenocarcinoma HeLa cells with methanolic extract of *P. santalinus* resulted in growth inhibition and induction of apoptosis with IC_{50} value 40 μg/ml. Methanolic extract of *P. santalinus* induced apoptosis through the mitochondrial pathway involving cytochrome c release from mitochondria, the activation of caspases 9 and 3 and degradation of peroxisome-activated receptor protein (PARP) (Kwon et al. 2006). Benzofuran compounds isolated from heartwood showed cytotoxicity against Ca9-22 cancer cells with an IC_{50} value 0.46 μg/ml (Wu et al. 2011a). The inhibitory activity of savinin on T-cell proliferation was highest from the methylene chloride extract (25 μg/ml) in comparison with other solvents ethyl acetate, n-butanol and water (Kwon et al. 2006).

Pterostilbene therapeutic efficacy was tested against breast cancer, lung cancer, colon cancer, prostate cancer and pancreatic cancer (Mena et al. 2012). Pterolinus K and pterolinus L isolated from heartwood have shown significant anticancer property with IC_{50} values at 10.86, 9.81 and 17–8.2 μM respectively against cancer cell lines HepG2, Hep3B and A549 (Wu et al. 2011b). Future studies are required using isolated compounds against various cancers and essential signalling pathways involved in the pathogenesis.

The leaf, stem and bark of *Pterocarpus santalinus* were evaluated by Donga et al. (2017) for their in vitro cytotoxicity potential by MTT assay. The plant parts were extracted by individual and successive cold percolation method using several solvents of varied polarity. The extractive yield varied with different solvents. The cytotoxic study of methanol extract of all the three parts extracted by both methods was done against HeLa cancer cell line, breast cancer cell line and normal cell line. All the three extracts showed dose-dependent cytotoxic effect; when compared, ME extract of stem showed minimum % cell viability, and among methods used, successive extraction methods were better than individual extraction method.

Though there was not much difference in % cell viability by employing either the method or by different parts. Hence it can be concluded that all the three parts showed cytotoxic effect against HeLa and breast cancer cell line (Donga et al. 2017).

6.12 Hepatoprotective Mechanisms

The use of natural remedies for the treatment of liver diseases has a long history, beginning with the Ayurvedic treatment and spreading to the Chinese, European as well as other systems of traditional medicines. Pharmacological validation of each hepatoprotective plant should include efficacy evaluation against liver diseases induced by various agents (Ding et al. 2012). *P. santalinus* bark and heartwood are rich in flavonoids and protect liver against chemical-induced toxicity. Studies showed that aqueous (45 mg/ml) and ethanol (30 mg/ml) bark extracts of *P. santalinus* restored CCl_4-induced liver damage in rats (Manjunatha 2006b). *P. santalinus* protects the liver from severe damage caused by D-GalN and may serve as a useful adjuvant in several clinical conditions associated with liver damage. Studies through in silico docking studies revealed that compounds present in *P. santalinus* heartwood, i.e. pterocarpol and cryptomeridiol targeted HBx proteins of hepatitis B virus and hence a strong drug candidate (Manjunatha et al. 2010). Itoh et al. (2009, 2010) reported that *P. santalinus* suppress hepatic fibrosis in chronic liver injury. Saradamma et al. (2016) demonstrated the protective effects of methanolic extract of *P. santalinus* heartwood against alcohol-induced hepatotoxicity. Also, researchers have to elucidate the underlying protective molecular mechanisms of isolated compounds against liver-related diseases.

6.13 Gastro-protective Mechanisms

P. santalinus contributes towards its marked gastro protection by its free radical scavenging property and also by maintaining the functional integrity of cell membrane through inhibition of acid secretion (Narayan et al. 2005). The anti-*Helicobacter pylori* activity of *P. santalinus* has been compared with that of bismuth subcitrate, through in vitro studies employing rat gastric epithelial cell cultures and *Helicobacter pylori* isolates from gastric mucosal biopsy patients. The use of ethanol extracts of *P. santalinus* as anti-*Helicobacter pylori* has been confirmed through its ability to reduce urease activity (Narayan et al. 2007a). Gastric mucosal injury results as a consequence of various conditions, including alcohol intake, refluxed bile salts, stress, aging, *Helicobacter pylori* infection and most of the non-steroidal anti-inflammatory drugs. Ethanol extract of *P. santalinus* was evaluated for gastro protection against ibuprofen-induced ulcers in albino rats. Previous reports revealed that extract of heartwood (150 mg/kg b.wt/day) showed a significant reduction in gastric lesions, increase in activities of antioxidant enzymes and decrease in membrane-bound ATPase activities and ability to maintain functional integrity of cell membranes (Narayan et al. 2005). Ethanolic extract of heartwood restored the activities of tricarboxylic acid (TCA) cycle enzymes, prevents mitochondrial dysfunction by providing mitochondrial membrane integrity and hydrophobic nature to the gastric mucosa and reverses the damage caused by ulcerogens at a dose of 200 mg/kg b.wt/day (Narayan et al. 2007b). The plant extract ability has not been studied against various chemical agents which disturb the gut microbial flora which are useful in physiology and also protect the intestinal mucosa.

6.14 Hypolipidaemic Activity

Generally, phenolic phytocompounds exhibit antioxidative, hypolipidaemic activities and thereby protects the liver and heart against a wide variety of pathological conditions including hyperlipidaemia by inhibiting lipid synthesis (Suanarunsawat et al. 2010). Modification of lipids is an integral component for initiation and progression of atherosclerosis in part by its effects on lipid profile. Circulatory VLDL, LDL and HDL are powerful markers and risk factors for cardiovascular disease (CVD). Lipid-lowering activity of bark extract (150 mg/kg b.wt/day) was studied in diabetic rats (Kondeti et al. 2010). Reduction in total cholesterol and lipoproteins and increase in HDL-C were observed in diabetic rats treated with aqueous extract of heartwood (250 mg/kg b.wt/day) in combination with vitamin E (Halim and Misra 2011). Chloroform extract of heartwood (400 mg/kg b. wt/day) significantly lowered the serum total cholesterol and triglycerides in D-galactosamine-intoxicated rats (Dhanabal et al. 2007). Pterostilbene has also been reported to have hypolipidaemic properties comparable to clinically used fibrate lipid-lowering drugs (Rimando et al. 2005). Riche et al. (2014) reported that pterostilbene increased LDL and reduced blood pressure in hypercholesterolemic human volunteers. The main etiologic factor in atherogenesis is oxidation of LDL. Reports showed that 3-hydroxybenzoic acid, gentisic acid, α- and β-resorcylic acid and vanillic acid present in heartwood of *P. santalinus* have potential inhibition of LDL oxidation (Ashidate et al. 2005). Plant extract is source for numerous complex compounds and not tested against protein factors which can regulate the expression of various transcription factors involved in enhanced lipid synthesis (Jadhav et al. 2012).

6.15 Angiogenesis and Wound Healing Activity

Process of generating new blood vessels from pre-existing ones is called angiogenesis. Angiogenesis process is controlled by positive and negative regulators. Chorio-allantoic membrane (CAM) assay is one of the well-established and commonly used in vitro models for evaluating angiogenesis and vasculogenesis. Jadhav et al. (2012) demonstrated that acetone extract of *P. santalinus* (1 mg/ml) stimulated angiogenesis more significantly than alcohol and benzene extracts. CAM angiogenic effect was mediated by the proliferation of endothelial cells without any toxicity (Jadhav et al. 2012). Oral administration (2 g/kg or 1 g/kg b.wt) and intraperitoneal injection

(500 mg/kg b.wt or 1 g/kg b. wt) were made to study the toxicity of the heartwood extract in rat models (Biswas et al. 2004). The plant also possesses the wound healing activity. Wound healing potentiality of the plant was demonstrated by Biswas et al. (2004) in normal and diabetic wound rat models and concluded that the ointment made from the plant is effective in treating the acute wounds. The angiogenic activity of plant may be attributed to the phytoconstituents present in the extract. Although the biological activities of specific compounds were not established, it is suggested that bioactive compounds present in extract might have acted by stimulating growth factors or signal cascade systems which needed further confirmation studies.

6.16 Other Pharmacological Uses

The heartwood extract showed strong analgesic action in mice by inhibiting the acetic acid-induced writhing by increasing the latency period (Achary and Ghaskadbi 2013). Phenolic acids in particular hydroxyl and polyhydroxy benzoic acids, 3-hydroxy benzoic acid, gentisic acid, α- and β-resorcylic acid and vanillic acid present in heartwood of *P. santalinus* were reported to have antifungal, antimutagenic, anti-inflammatory (Ashidate et al. 2005) and nematicidal activity (Sultana et al. 2010) and thyroid peroxidase inhibitory effect (Chattopadhyay et al. 2003). Many other reports also have shown that lignans possess antiviral, antiallodynic and antimutagenic properties (Reddy et al. 2006; Yamauchi et al. 2005; Kayser et al. 1998). Aurone glycosides isolated from heartwood have been reported to exhibit leishmanicidal activity against *Leishmania donovani*, *L. enrietti* and *L. major* (Kayser et al. 2001). And also, it was reported to inhibit in vitro erythrocytic stages of *Plasmodium falciparum* strains (Haschek et al. 2010). Traditional system of medicine suggested *P. santalinus* could be used to treat various medical complications like haemorrhage, dysentery, aphrodisiac, diaphoretic, eye diseases and mental aberrations (Kirtikar and Basu 2012). It is also used in treating leprosy, bone fracture, scorpion sting, hiccough and snake bites (Arokiyaraj et al. 2008). Freshly prepared red sanders wood paste applied on the face gives the positive reaction on itchy erythema and mild oedema because of santalin A and santalin B (Sandra et al. 1996). Methanolic extract of *P. santalinus* showed significant protection against alcohol-induced nephrotoxicity in rats (Bulle et al. 2016). Red sanders wood powders with following individual extracts have shown specific activity. Benzene extract showed antiviral activity against *encephalomyocarditis* (Shankaranarayana and Theagarajan 2000). Studies revealed the functional role of santalin against tyrosinase, in which santalin acts as inhibitor of tyrosine by interacting with the fluorophore amino acid residue of tyrosinase (Hridya et al. 2016). Further, heartwood ethanolic extract of *P. santalinus* regulated UVB irradiation-induced procollagen reduction and matrix metalloproteinases expression through activation of TGF-β/Smad and inhibition of the MAPK/AP-1 pathway in normal human dermal fibroblasts (Gao et al. 2017).

6.17 Pharmacological Activities of Individual Phytocompounds Isolated from *P. santalinus* (Soundararajan et al. 2016)

6.17.1 β-Amyrin

β-Amyrin exhibits various important pharmacological/biological activities such as analgesic, antiedemic, anti-inflammatory, antinociceptive, antiulcer, gastroprotective, hepatoprotective, larvicide and mosquitocide.

6.17.2 β-Eudesmol

β-Eudesmol exhibits various important pharmacological/biological activities such as antianoxic, antimutagenic, antipeptic, anti-*Salmonella*, antitumour promoter, antiulcer, central nervous system (CNS) inhibitor, calcium antagonist, hepatoprotective, neurogenic and sedative.

6.17.3 β-Sitosterol

β-Sitosterol exhibits various important pharmacological/biological activities such as androgenic, angiogenic, anorexic, antiadenomic, antiandrogenic, antibacterial, anticancer (breast), anticancer (cervix, lung), antiedemic, antioestrogenic, antifeedant, antifertility, antigonadotrophic, anti-hyperlipoproteinaemic, anti-inflammatory, antileukaemic, antilymphomic, antimutagenic, antiophidic, antioxidant, antiprogestational, antiprostaglandin, antiprostatadenomic, antiprostatitic, antipyretic, antitumour (breast), antitumour (cervix, lung), antiviral, apoptotic, artemicide, cancer-preventive, candidicide, caspase-8 inducer, oestrogenic, febrifuge, gonadotrophic, hepatoprotective, hypocholesterolaemic, hypoglycaemic, hypolipidaemic, spermicide, ubiquiot and ulcerogenic.

6.17.4 Betulin

Betulin exhibits various important pharmacological/biological activities such as anti-HIV, anti-carcinomic, antifeedant, antiflu, anti-inflammatory, antitumour, antiviral, aphidifuge, cytotoxic, hypolipemic, prostaglandin synthesis inhibitor and topoisomerase II inhibitor.

6.17.5 Cedrol

Cedrol exhibits various important pharmacological/biological activities such as termiticide.

6.17.6 Epilupeol

Epilupeol exhibits various important pharmacological/biological activities such as antiviral.

6.17.7 Gallic Acid

Gallic acid exhibits various important pharmacological/biological activities such as an angiotensin-converting enzyme inhibitor, analgesic, anti-HIV, anti-MRSA, anti-adenovirus, antiallergenic, anti-anaphylactic, anti-angiogenic, antiasthmatic, antibacterial, anti-bronchitic, anticancer, anti-carcinomic, anti-*Escherichia*, anti-fibrinolytic, antiflu, anti-hepatotoxic, anti-herpetic, anti-inflammatory, anti-leishmanic, antimutagenic, anti-nitrosaminic, antioxidant, antiperiodontitic, anti-peroxidant, anti-polio, anti-proteolytic, anti-radicular, antiseptic, anti-*Staphylococcus*, antitumour, antitumour promoter, antiviral, apoptotic, astringent, bacteristat, bronchodilator, cancer-preventive, candidicide, carcinogenic choleretic, cyclooxygenase inhibitor, cytotoxic, floral inhibitor, Gram(+)icide, Gram(−)icide, haemostat, hepatoprotective, immunomodulator, immunostimulant, Immunosuppressant, insulin-sparing, myorelaxant, NO inhibitor, nephrotoxic, styptic, topoisomerase I inhibitor and xanthine oxidase inhibitor.

6.17.8 Liquiritigenin

Liquiritigenin exhibits various important pharmacological/biological activities such as antidepressant, anti-inflammatory, anti-leukaemic, anti-spasmodic, antiulcer, central nervous system (CNS) active, cancer-preventive, fungicide, haemoglobin inducer, monoamine oxidase inhibitor and phytoalexin.

6.17.9 Lupeol

Lupeol exhibits various important pharmacological/biological activities such as anti-EBV (Epstein-Barr virus), anti-angiogenic, anti-oedemic, antiflu, anti-hyperglycaemic, anti-inflammatory, antilithic, antimalarial, anti-oxalate, antioxidant, anti-peroxidant, anti-prostaglandin, anti-rheumatic, antitumour, anti-urethrotic, antiviral, cytotoxic, farnesyl protein transferase (FPTase) inhibitor, hypotensive and topoisomerase (TOPO) II inhibitor.

6.17.10 Pterostilbene

Pterostilbene exhibits various important pharmacological/biological activities such as antibacterial, antidiabetic, anti-melanomic, antioxidant, anti-radicular, cyclooxygenase (COX) 1 inhibitor, COX 2 inhibitor, chemopreventive, cyclooxygenase inhibitor, cytotoxic, fungicide, hyperglycaemic, hypocholesterolemic, hypoglycaemic, hypolipidaemic, hypotensive, insecticide and phytoalexin.

6.18 Perspectives of Future Research

From compositional point of view, the phytocompounds present in heartwood of *P. santalinus* are very complex compounds. Many compounds which were identified are general preliminary compounds, but each preliminary compound is a mixture of many more compounds which are not identified as therapeutic agents. Further research is necessary to provide a detailed characterization of the compounds present in the preliminary compounds. Limited studies were suggested that radical scavenging and beneficial effects of heartwood of *P. santalinus* and the molecular mechanisms of the biological activities and physiological effects need further study. Also the bioavailability and metabolism of the compounds have not been reported. Further studies on these aspects would provide useful data for substantiating the physiological effects of these compounds. Most of the biological activities and physiological effects of the plant extracts have been studied in vitro or in animal models. Biological activities have not been reported for the following individual chemical component of red sanders: acetyl-oleanolic acid; acetyloleanolaldehyde; beta-amyrone; cryptomeridiol; desoxysantalin; erythrodiol; eudes-4(15)-ene-2, 11-diol; homopterocarpin; isopterocarpin; isopterocarpolone; Lup-20(29)-ene-2alpha, 3beta-diol; lupenone; pterocarpdiolone; pterocarpin; pterocarpol; pterocarptriol; santal; and santalin (A and B) (Duke's 2015). Clinical and human intervention studies are very limited; therefore it is also worth to investigate the biological and physiological effects of the isolated compounds of the heartwood.

References

Achary JD, Ghaskadbi SS (2013) Protective effect of pterostilbene against free radical mediated oxidative damage. BMC Complement Altern Med 13:238

Arokiyaraj A, Perinbam K (2010) Antifungal activity of *Pterocarpus santalinus* an in vitro study. Biomed Pharmacol J 3(1):263–265

Arokiyaraj S, Martin S, Perinbam K, Marie AP, Beatrice V (2008) Free radical scavenging activity and HPTLC fingerprint *Pterocarpus santalinus* L.-an in vitro study. Indian J Sci Technol 1:1–3

Arunakumara KKIU, Subasinghe BCWS, Yoon MH (2011) *Pterocarpus santalinus* Linn. f. (Roth handgun): a review of its botany, uses, phytochemistry and pharmacology. J Korean Soc Appl Biol Chem 54:495–500

Ashidate K, Kawamura M, Mimura D, Tohda H, Miyazaki S et al (2005) Gentisic acid, an aspirin metabolite, inhibits oxidation of low-density lipoprotein and the formation of cholesterol ester hydroperoxides in human plasma. Eur J Pharmacol 513:173–179

Azamthulla M, Rajkapoor B, Kavimani S (2015) A review on *Pterocarpus santalinus* Linn. f. World J Pharm Res 4(2):282–292

Balaraju K, Arokiyaraj S, Agastian P, Perinbam K, Ignacimuthu S (2008) Antimicrobial activity of leaf extracts of *Pterocarpus santalinus* L. (Fabaceae). J Pure Appl Microbiol 2(1):161–164

Bhagyaraj A (2017) Red Sanders invisible activities in Chittoor district: a case study. Int J Adv Res 5(9):1073–1080

Biswas TK, Narayan ML, Mukherjee B (2004) Wound healing potential of *Pterocarpus santalinus* Linn.f.: a pharmacological evaluation. Int J Low Extrem Wounds 3:143–150

Bulle S, Reddy VD, Hebbani AV, Padmavathi P, Challa C, Puvvada PK, Repalle E, Nayakanti D, Narasimhulu CA, Nallanchakravarthula V (2016) Nephro-protective action of *P. santalinus* against alcohol-induced biochemical alterations and oxidative damage in rats. Biomed Pharmacother 84:740–746

Chattopadhyay SK, Kumar TRS, Maulik PR, Srivastava S, Garga A (2003) Absolute configuration and anticancer activity of taxiresinol and related lignans of *Taxus wallichiana*. Bioorg Med Chem 11:4945–4948

Cho JY, Park J, Kim PS, Yoo ES, Baik KU, Park MH (2001) Savinin a lignin from *Pterocarpus santalinus* inhibits tumor necrosis factor-alpha production and T-cell proliferation. Biol Pharm Bull 24:167–174

Dey SK, Chattopadhay S, Masanta NC (2014) Antimicrobial activities of some medicinal plants of red and laterite zone of West Bengal, India. World J Pharm Pharm Sci 3:719–734

Dhanabal P, Kannan SE, Bhojraj S (2007) Protective and therapeutic effects of the Indian medicinal plant *Pterocarpus santalinus* on D-galactosamine-induced liver damage. Asian J Trad Med 2:51–57

Dhande PP, Gupta AO, Jain S, Dawane JS (2017) Anti-inflammatory and analgesic activities of topical formulations of *Pterocarpus santalinus* powder in rat model of chronic inflammation. J Clin Diagn Res 11(7):FF01–FF04

Ding RB, Tian K, He CW, Jiang Y, Wang YT, Wan JB (2012) Herbal medicines for the prevention of alcoholic liver diseases. J Ethn Pharm 144:457–465

Donald PB (2000) Medicinal plants and phytomedicines. Linking plant biochemistry and physiology to human health plant. J Physiol 124:507–514

Donga S, Moteriya P, Chanda S (2017a) Evaluation of antimicrobial and synergistric antimicrobial properties of *Pterocarpus santalinus*. Asian J Pharm Clin Res 10(11):204–209

Donga S, Pande J, Moteriya P, Chanda S (2017b) In vitro cytotoxicity of leaf, stem, bark of *Pterocarpus santalinus* Linn.f. Phamacogn. Phytochemistry 6(4):297–305

Duke's Phytochemical and Ethno Botanical Databases. http://www.ars-grin.gov/cgi-bin/duke/farmacy2.pl. Viewed 8th June 2015

Faremi TY, Suru SM, Fafunso MA, Obioha UE (2008) Hepato-protective potentials of *Phyllanthus amarus* against ethanol-induced oxidative stress in rats. Food Chem Toxicol 46:2658–2664

Gao W, Lin P, Hwang E, Wang Y, Yan Z, Ngo HTT, Yi TH (2017) *Pterocarpus santalinus* L. regulated ultraviolet B irradiation-induced procollagen reduction and matrix metalloproteinases expression through activation of TGF-β/Smad and inhibition of the MAPK/AP-1 pathway in normal human dermal fibroblasts. Photochem Photobiol. https://doi.org/10.1111/php.12835. [Epub ahead of print]

Gopinath K, Gowri S, Arumugam A (2013) Phytosynthesis of silver nanoparticles using *Pterocarpus santalinus* leaf extract and their antibacterial properties. J Nanostruct Chem 3:68–85

Halim ME, Misra A (2011) The effect of aqueous extract of *Pterocarpus santalinus* heartwood and vitamin E supplementation in streptozotocin-induced diabetic rats. J Med Plants Res 5:398–409

Haschek WM, Wallig MA, Rousseaux C (2010) Fundamentals of toxicologic pathology. Academic, San Diego

Hridya H, Amrita A, Mohan S, Gopalakrishnan M, Dakshinamurthy TK, Doss GP, Siva R (2016) Functionality study of santalin as tyrosinase inhibitor: a potential depigmentation agent. Photochem Photobiol 86:383–389

Huang WY, Cai YZ, Zhang Y (2009) Natural phenolic compounds from medicinal herbs and dietary plants: potential use for cancer prevention. Nutr Cancer 62(1):1–20. https://doi.org/10.1080/01635580903191585

Itoh A, Isoda K, Kondoh M, Kawase M, Kobayashi M, Tamesada M, Yagi K (2009) Hepatoprotective effect of syringic acid and vanillic acid on concanavalin A-induced liver injury. Biol Pharm Bull 32:1215–1219

Itoh A, Isoda K, Kondoh M, Kawase M, Watari A, Kobayashi M, Tamesada M, Yagi K (2010) Hepatoprotective effect of syringic acid and vanillic acid on CCl_4-induced liver injury. Biol Pharm Bull 33:983–987

Jadhav J, Mane A, Kanase A (2012) Stimulatory effect of *Pterocarpus santalinus* on vasculogenesis in chick chorioallantoic membrane (CAM). J Pharm Res 5:208–211

Kayser O, Kiderlen AF, Folkens U, Kolodziej H (1998) In vitro leishmanicidal activity of aurones. Planta Med 65:316–319

Kayser O, Kiderlen AF, Brun B (2001) In vitro activity of aurones against *Plasmodium falciparum* strains K1 and NF54. Planta Med 67:718–721

Kirtikar KR, Basu BD (2012) Indian medicinal plants. Biogreen Books, New Delhi

Kondeti VK, Rao BK, Maddirala DR, Thur SKM, Fatima SS, Kasetti RB, Rao CA (2010) Effect of *Pterocarpus santalinus* bark, on blood glucose, serum lipids, plasma insulin and hepatic carbohydrate metabolic enzymes in streptozotocin-induced diabetic rats. Food Chem Toxicol 48:1281–1287

Krishnamoorthy P, Stella J, Mohamed AJ, Anand M (2011) Radical scavenging and antibacterial evaluation of *Pterocarpus santalinus* leaf in-vitro study. Int J Pharmacol Sci Res 2:1204–1208

Krishnaveni KS, Srinivasa Rao JV (2000) A new isoflavone glucoside from *Pterocarpus santalinus*. J Asian Nat Prod Res 2(3):219–223

Kumar D (2011) Anti-inflammatory, analgesic, and antioxidant activities of methanolic wood extract of *Pterocarpus santalinus* L. J Pharmacol Pharmacother 2:200–202

Kwon HJ, Hong YK, Kim KH, Han CH, Cho SH, Choi JS, Kim BW (2006) Methanolic extract of *Pterocarpus santalinus* induces apoptosis in Hela cells. J Ethnopharmacol 105:229–234

Lokesh T, Sarada B, Swarnalatha K, Kareem MA, Varadacharyulu NC (2016) Antihaemolytic activity of phytochemical aqueous extracts of *Pterocarpus santalinus* and *Phyllanthus emblica* in red blood cells of human subjects receiving chronic alcohol and cigarette smoking. Int J Pharm Sci Res 7(9):3857–3864

Lopez AD, Mathers CD, Ezzati M, Jamison DT, Murray CJ (2006) Global burden of disease and risk factors. Oxford University Press, New York

Manjunatha BK (2006a) Antibacterial activity of *Pterocarpus santalinus*. Indian J Pharm Sci 68:115–116

Manjunatha BK (2006b) Hepatoprotective activity of *Pterocarpus santalinus* L. f., an endangered plant. Indian J Pharmacol 38:25–28

Manjunatha BK, Amit R, Priyadarshini P, Paul K (2010) Lead findings from *Pterocarpus santalinus* with hepatoprotective potentials through in silico methods. Int J Pharma Sci Res 7:265–270

Mena S, Rodríguez ML, Ponsoda X, Estrela JM, Jatteela M et al (2012) Pterostilbene-induced tumor cytotoxicity: a lysosomal membrane permeabilization-dependent mechanism. PLoS One 7:e44524

Mishra MP, Padhy RN (2013) In Vitro antibacterial efficacy of 21 Indian timber-yielding plants against multidrug-resistant bacteria causing urinary tract infection. Osong Pub Health Res Perspect 4:347–357

Mohandass G, Devi KS, Devanesan (2010) Antifungal activity of *Pterocarpus santalinus*, an in vitro study. Biomed Pharmacol J 3(1):107–110

Nagaraju N, Rao KN (1989) Folk medicine for diabetes from Rayalaseema of Andhra Pradesh. Anc Sci Life 9:31–35

Nagaraju N, Prasad M, Gopalakrishna G, Rao KN (1991) Blood sugar lowering effect of *Pterocarpus santalinus* (Red Sanders) wood extract in different rat models. Int J Pharmacogn 29:141–144

Narayan S, Devi RS, Srinivasan P, Devi CSS (2005) *Pterocarpus santalinus*: a traditional herbal drug as a protectant against ibuprofen induced gastric ulcers. Phytother Res 19:958–962

Narayan S, Veeraraghavan M, Devi CS (2007a) *Pterocarpus santalinus*: an in vitro study on its anti-*Helicobacter pylori* effect. Phytother Res 21:190–193

Narayan S, Devi RS, Shyamala Devi CS (2007b) Role of *Pterocarpus santalinus* against mitochondrial dysfunction and membrane lipid changes induced by ulcerogens in rat gastric mucosa. Chem Biol Interact 170:67–75

Padmalatha K, Prasad MNV (2006) *Pterocarpus santalinus* L.f. and *P. marsupium* Roxb. – potential source of phytomedicines. In: Sharma RK, Arora R (eds) Herbal drugs: a twenty first century perspectives. Jaypee Brothers, Medicinal Publishers, New Delhi, pp 450–459

Panayi GS, Lanchbury JS, Kingsley GH (1992) The importance of the T cell in initiating and maintaining the chronic synovitis of rheumatoid arthritis. Arthritis Rheum 35:729–735

Parrotta JA (2001) Healing plants of peninsular India. CABI Publications, Wallingford

Prasad SJ, Menon JM, Vijil VV, Aswathi PS, Nimitha KN, Athira L, Vijayalakshmi SV, Sruthi PS, Sughi S, Kurien NK (2016) Evaluation of anti-melanogenic activity if *Pterocarpus santalinus* L. using bacterial system. Biotech. Res 2(2):58–60

Pullaiah T (2006) Encyclopaedia of world medicinal plants. Regency Publications, New Delhi

Ramabrahmam V, Sujatha (2016) Red Sanders in Rayalaseema region of Andhra Pradesh: importance to commercial & medicinal value. IOSR J Pharm Biol Sci 11(1):57–60

Rao KB, Giri R, Kesavulu MM, Apparao C (2001) Effect of oral administration of bark extracts of *Pterocarpus santalinus* on blood glucose level in experimental animals. J Ethnopharmacol 74:69–74

Reddy OSK (2018) Red Sanders – *Pterocarpus santalinus* – Rakta chandan – The Red Gold – Green wealth – An opportunity for Indian Farming community – Let's explore. https://www.linkedin.com/pulse/red-sanders-gold-green-wealth-future-generations-indian-osk-reddy/

Reddy L, Reddy G, Reddy RS, Reddy S, Reddy J (2006) Characterization of red sandalwood by ICP-MS analysis, optical absorption and EPR spectroscopic methods. J Rad Eff Defects Solids 161(9):537–547

Riche DM, Riche KD, Blackshear CT et al (2014) Pterostilbene on metabolic parameters: a randomized, double-blind, and placebo-controlled trial. J Evid Based Complement Altern Med. https://doi.org/10.1155/2014/459165

Rimando AM, Nagmani R, Feller DR, Yokoyama W (2005) Pterostilbene, a new agonist for the peroxisome proliferator-activated receptor alpha-isoform, lowers plasma lipoproteins and cholesterol in hypercholesterolemic hamsters. J Agric Food Chem 53:3403–3407

Sandra A, Shenoi SD, Srinivas CR (1996) Allergic contact dermatitis from red sandalwood (*Pterocarpus santalinus*). Contact Dermatitis 34(1):69

Saradamma B, Reddy VD, Padmavathi P, Paramahamsa M, Varadacharyulu N (2016) Modulatory role of *Pterocarpus santalinus* against alcohol-induced liver oxidative/nitrosative damage in rats. Biomed Pharmacother 83:1057–1063

Shankaranarayana KH, Theagarajan KS (2000) Antimicrobial properties of red sanders extracts. Wood News 8(4):25–26

Soundararajan V, Ravi Kumar G, Murugesan K, Chandrashekar BS (2016) A review on red sanders (*Pterocarpus santalinus* Linn.f.)-Phytochemistry and pharmacological importance. World J Pharm Pharm Sci 5(6):667–689

Suanarunsawat T, Boonnak T, Ayutthaya WD, Thirawarapan S (2010) Anti-hyperlipidemic and cardioprotective effects of *Ocimum sanctum* L. fixed oil in rats fed a high fat diet. J Basic Clin Physiol Pharmacol 2:387–400

Sultana N, Akhter M, Khatoon Z (2010) Nematicidal natural products from the aerial parts of *Rubus niveus*. Nat Prod Res 24:407–415

Ullah MF, Khan MW (2008) Food as medicine: potential therapeutic tendencies of plant derived polyphenolic compounds. Asian Pac J Cancer Prev 9:187–195

Vandita P, Amin N, Khyati P, Monisha K (2013) Effect of phytochemical constituents of *Ricinus communis*, *Pterocarpus santalinus*, *Terminalia bellerica* on antibacterial, antifungal and cytotoxic activity. Int J Toxicol Pharm Res 5(2):47–54

Wu SF, Chang FR, Wang SY, Hwang TL, Lee CL, Chen SL, Wu CC, Wu YC (2011a) Anti-inflammatory and cytotoxic neoflavonoids and benzofurans from *Pterocarpus santalinus*. J Nat Prod 74:989–996

Wu SF, Chang FR, Wang SY, Hwang TL, Lee CL, Chen SL, Wu CC, Wu YC (2011b) Bioactive components from the heartwood of *Pterocarpus santalinus*. J Bioorg Med Chem Lett 21:5630–5632

Yamauchi S, Hayashi Y, Nakashima Y, Kirikihira T, Yamada K, Masuda T (2005) Effect of benzylic oxygen on the antioxidant activity of phenolic lignans. J Nat Prod 68:1459–1470

Reproductive Biology 7

T. Pullaiah

Abstract

In this chapter pollination ecology of red sanders has been described. In flowering phenology, pollination mechanism has been given in detail. The endemic nature to certain extent is related to floral biology of this tree, and this information contributes to conservation and tree improvement programmes. Pollination is only by honey bee species *Apis dorsata*, *A. cerana indica* and *A. florae*. In red sanders fruit formation is by xenogamy, autogamy and geitonogamy, which indicates that breeding system in this species is facultative xenogamous. Compared to very high flower production, the fruit set is very low.

Keywords

Flowering phenology · Floral biology · Pollination ecology · Pollination mechnaism · Xenogamy

7.1 Introduction

Red Sanders, colloquially called as Rakta Chandan, is an important tree species and a matter of concern to ecologists, forest scientists and environmentalists for several decades. Restricted distribution patterns, long juvenile period and tough access to mature flowering trees may be reasons for its limited study, and floral biology is one such grey area, and hardly a few publications are witnessed. To undertake a breeding programme and efficiently manage seed orchards or maximize seed production in seed production areas, the reproductive phenology, pollination biology, seed and fruit ontogeny and constraints to seed and fruit production must be understood. The main challenge of a breeder in this species is to provide wavy-grained saplings to the

T. Pullaiah (✉)
Department of Botany, Sri Krishnadevaraya University, Anantapur, Andhra Pradesh, India

T. Pullaiah et al. (eds.), *Red Sanders: Silviculture and Conservation*,
https://doi.org/10.1007/978-981-13-7627-6_7

farmers, to ignite sustainable utilization. As a member of Fabaceae, this is a cross-pollinator, and the information on pollination biology is critical for propagating the trees with wavy-grained wood, considered to be the main trait in terms of wood quality.

7.2 Pollination Ecology

Plant-pollinator interactions are important for the maintenance of biodiversity and are one of the critical services for sustainable ecosystems. However, studies point that landscape changes resulting from anthropogenic disturbance, as habitat loss, fragmentation and degradation, are one of the primary threats to pollination services (Pinheiro et al. 2018). *Pterocarpus* is a tropical genus consisting of 46 species, 5 of which occur in the Indo-Pacific region. Most of the species are found in Western Tropical Africa. Of these 46 species, reproductive biology has been studied only in 5 species – *Pterocarpus indicus* (Escobin et al. 2004), *P. santalinus* (Rao et al. 2001; Rao and Raju 2002), *P. macrocarpus* (Doungyotha and Owens 2002), *P. marsupium* (Pal and Mondal 2018) and *P. violaceus* (Pinheiro et al. 2018). Rao et al. (2001) and Rao and Raju (2002) studied flowering phenology, pollen viability, breeding behaviour and pollination ecology of *Pterocarpus santalinus*. The following account is mainly based on their studies.

7.2.1 Flowering Phenology

The trees exhibited leaf fall from January to middle of March and new foliage appearing in April. Flowers were born in late March and continued up to late May along with the new foliage. The proportion of anthesized flowers increased gradually in number and reached a peak in 2nd and 3rd week of April and then decreased gradually to terminate in the 2nd/3rd week of May. Flowering at population level was asynchronous. The duration of flowering period ranged between 8 and 32 days, i.e. with an average of 20 days. During this period, mass blooming is seen on certain days, but there is a gap between 2 mass flowering days (Rao et al. 2001).

Janzen (1967) and Frankie et al. (1983) are of the opinion that most tropical tree species that flower in the dry season are mass blooming, and *Pterocarpus santalinus* is in agreement with this generalization. Interestingly, in red sanders mass bloom recurs only on certain days with definite period of interval and does not produce mass bloom continuously during the flowering season. Such intermittent mass blooming appears to be unique as no such pattern is found in the descriptions of flowering patterns of tropical plants. This pattern of flowering appears to be a consequence of the dry arid climate with temperatures reaching a high of 45 °C and resulting in water stress. It is likely that water and energy are conserved in the period between mass blooms and invested to produce the subsequent mass bloom and so on (Rao et al. 2001).

7.2.2 Inflorescence Phenology

Flowers are in a simple, *c.* 7 cm long, branched raceme and consist of 25 ± 4 flowers which dehisce intermittently over a period of 15 days.

7.2.3 Flower Morphology

The flowers are bright yellow, 16 mm long, typically papilionaceous, zygomorphic, bisexual and emitted mild fragrance and pedicellate, pedicel to 5 mm long. Sepals five, green, joined at the base to form a tube of 6–7 mm in length. Petals 5, standard petal posterior, ovate with crisped margins, *c.* 2.1 × 1.5 cm; lateral two winged petals *c.* 1.8 × 0.8 cm; the two keel petals are loosely contiguous and *c.* 1.4 × 0.35 cm. The gynoecium/androecium set is concealed within the keel petals. Stamens are monadelphous with ten anthers and anthers are dithecous. Rao and Raju (2002) have erroneously mentioned that stamens are united into two bundles of five each. Unilocular superior ovary contains two ovules. In bud stage, the standard petal encloses the wing and keel petals which in turn enclose the sex organs. As the bud matures, the standard petal gradually bulges and protrudes from the calyx. Pinheiro et al. (2018) mentioned that in *P. violaceus*, the style and the staminal tube follow the keel's curvature upwards, and anthers and stigma are positioned near the keel apex. The same condition is met with in all the species of *Pterocarpus* including *P. santalinus*. They also reported that nectar is the main source, accumulated around the base of the gynoecium and confined within the chamber formed by the fusion of the filaments, and can be reached only by small openings on the base of the upper side of the staminal tube, whose access is blocked by the claw of the standard petal.

7.2.4 Anthesis

Flowers opened from 0030 h in acropetal order; they unfold exposing the wing and keel petals in which the sex organs are still enclosed. The flowers secrete nectar in trace amount. Anther dehiscence took place asynchronously; two anthers an hour after anthesis, three anthers again after 1 h and the other five again after 1.5 h. Anther dehiscence occurred by longitudinal slits. During the period of anther dehiscence, the temperature ranged between 30 and 33 °C and RH between 74% and 78%. The pollen grains are fertile, 21 mm in size, yellow in colour, spherical, tricolpate and smooth-walled. Each anther produces a mean number of 3480 pollen grains per anther, and each flower produces around 34,800 pollen grains accordingly. The ratio of pollen grain to ovule number is 17,400: 1. The pollen grains germinated well in a 70% sucrose solution. Fresh pollen germinated to the extent of 88%. The germinability of the pollen stored for 12 h decreased to 60%, those stored for 24 h to 32% and those stored for 36 h to 13%. The pollen stored further did not germinate (Rao et al. 2001; Rao and Raju 2002).

7.2.5 Pollination Biology

Successful pollination is determined by the flower-opening time, anther dehiscence and also stigma receptivity. The stigma becomes receptive by about the time the first two anthers dehisce and remains so until late evening of the same day (Rao and Raju 2002). The hand-pollination tests for stigma receptivity showed 84% fruit set with fresh stigmas and 8% with 18-h-old stigmas. The hand-pollination tests for breeding systems indicated 24% fruit set and 50% seed set through autogamy, 68% fruit set and 50% seed set through geitonogamy and 84% fruit set and 57% seed set through xenogamy. Of the two ovules per flower, only one ovule produced seed in all the modes of pollination, but rarely both ovules produced seeds in xenogamous fruits. Most of the autogamous and geitonogamous fruits dropped off prematurely, while all xenogamous fruits were retained to maturity (Table 7.1). The natural fruit set was 6%. A sample of 100 inflorescences consisting of 2646 flowers with 5292 ovules selected at random on different trees at flower age was used for estimating fruit set, seed set and fecundity. Among these are 170 flowers with 340 ovules set fruits with 178 seeds. Seed set was 52%. Fecundity was 3%, which was expressed in terms of total number of seeds produced against the total number of ovules in sampled inflorescences (Rao and Raju 2002).

The flowers were exclusively foraged by honey bees *Apis dorsata* (rock bee), *A. cerana indica* and *A. florae*. Bee flowers are zygomorphic, mechanically strong having landing platform and semiclosed petals. According to Endress (1995), bees prefer yellow flowers. Among the foraging honey bees, rock bee was the most dominant in visits and number and visited soon after anthesis on moonlit nights, showing brisk activity up to 0600 h. Its activity later declined and disappeared at around 0730 h. This may be due to the high temperature during these summer months. *Bauhinia racemosa* was the co-flowering tree in the same habitat, and it anthesed during 1530–1830 h. *Apis dorsata* gathered pollen from this plant during moonlit hours alternately. *A. cerana indica* and *A. florae* visited the flowers between 0530 and 0740 h only. These two species were relatively less in number and visits. Papilionoids are pollinated mainly by large- and medium-sized bee species, and red sanders pollination is in consonance with this statement. *A. dorsata* is the largest among the bee species and hence effective in pollination. Both nectar and pollen

Table 7.1 Results of breeding systems in *P. santalinus* (After Rao and Raju 2002)

Treatment	No. of flowers pollinated	No. of flowers set fruit	Fruit set (%)	No. of seeds produced	Seed set (%)	No. of fruits dropped off prematurely	Fruit drop (%)
Apomixis	50	0	0	0	0	0	0
Autogamy unmanipulated	50	0	0	0	0	0	0
Autogamy manipulated	50	12	24	12	50	10	83
Geitonogamy	50	34	68	34	50	20	59
Xenogamy	50	42	84	48	57	0	0

Fig. 7.1 Pollinator bees of *P. santalinus*. (**a**) *Apis dorsata*, (**b**) *A. cerana indica* and (**c**) *A. florae*. (After Rao and Raju 2002)

constitute rewards for the flower visitors. The stigma is slightly curved upwards and extended beyond the length of the stamens. Both stamens and pistils are enclosed by the wing and keel petals in open flowers. The honey bees collected both pollen and nectar indiscriminately (Fig. 7.1).

Bees landed on the corolla wing-keel complex and directed their heads between the claw of the flag and basal portion of the staminal tube where they extended the proboscis through small openings to reach the nectar. Thus, the flower keel would move down, exposing anthers and stigma, which then touch the ventral side of the bee's abdomen. After the bee departed, the petals would return to their original position. In effect, the bees picked up pollen on their underside, and this pollen gets transferred to other conspecific stigmas in their subsequent visits. On mass blooming days, the honey bees mostly concentrated on the same tree which they first visited for collecting floral rewards, while on days of little flowering, they frequently visited the flowers of conspecific trees for want of floral rewards. *A. dorsata* was efficient in harvesting floral rewards by spending very little time per flower. The other two honey bee species were relatively inefficient in depleting floral rewards as they spent more time per flower (Table 7.2). The body washings of honey bees for pollen grains indicated that *A. dorsata* carried more pollen than the other two honey bee species. The difference between mean number of pollen grains carried by *A. dorsata* and *A. cerana indica* is just significant and that between *A. dorsata* and *A. florae* is highly significant. Therefore, all the three honey bee species were pollen carriers and affected pollination. However, *A. dorsata* was the main pollinator, especially for cross-pollination because of its frequent inter-tree movements (Rao and Raju 2002).

P. santalinus fruits through autogamy, geitonogamy and xenogamy, indicating that it has facultative xenogamous breeding system. Autogamy does not take place without the pollen mediation by the flower visitors. This is because the automatic contact between the anthers and the stigma within the same flower does not take place as the stigma is placed well beyond the anthers. The same is realized in the hand-pollination test for unmanipulated autogamy. Therefore, both self- and cross-pollinations depend on pollen mediation performed by the flower visitors. Although red sanders is both self- and cross-pollinating, it sheds most of the autogamous and

Table 7.2 Foraging efficiency and pollen pick-up by honeybees on *P. santalinus* (After Rao and Raju 2002)

Visitor species	No. of flowers visited				Length of a visit				Pollen grains found in body washings			
	N	*R*	$\bar{X}$	*SD*	*N*	*R*	$\bar{X}$	*SD*	*N*	*R*	$\bar{X}$	*SD*
Apis dorsata	10	12–18	15	2.56	10	1–6	3	1.4	10	850–1860	1346	365
A. cerana indica	10	4–14	8	3.17	10	2–13	6	3.0	10	472–1420	914	530
A. florae	10	6–12	9	1.92	10	4–15	7	3.9	10	340–985	726	240

N No. of insects observed, *R* range, *X* mean, *SD* standard deviation geitonogamous fruits, while retaining all xenogamous fruits to maturity. It suggests that *P. santalinus*, by predominantly cross-pollinating, leaves open the possibilities for self-pollination. This tree species might be selectively eliminating the growing self-pollinated offspring to allocate resources for the xenogamous fruits. The pollen-ovule ratio found in *P. santalinus* is much more than that predicted for facultative xenogamy. This high pollen-ovule ratio appears to be imperative for the success of facultative xenogamy, especially for xenogamy (Rao and Raju 2002).

Compared to very high flower production, the natural fruit set is very low in red sanders. Different factors could affect low fruit set. First, this tree being able to fruit through both self- and cross-pollination initiates more fruit production but gradually and selectively eliminates the growing poor offspring, especially those resulting from self-pollination. The compatibility to self-pollinate seems to provide fertility assurance in the event of failure of outcrossing. Secondly, this species is expected to maintain lower levels of variation because of its endemic status with restricted population size, and consequently it is reducing opportunities for outcrossing. Further, it is endangered due to constant anthropogenic pressure, and any reduction in the population size is bound to enforce inbreeding and genetic bottlenecks. Thirdly, the low maternal energy investment is available during dry period for the rapidly growing offspring. These factors might be collectively responsible for the low natural fruit set rate (Rao and Raju 2002).

The pollination ecology of *P. santalinus* is vulnerable to dry and hot conditions. The flowers remain unvisited day-long because of the absence of pollinator activity. The natural fruit set is a consequence of pollinator activity during moonlit night and early morning hours. In red sanders fruit formation through xenogamy is better than fruit formation through geitonogamy and autogamy. Therefore, the breeding system is mixed with self-pollination and outcrossing. In the mating system, outcrossing is predominant, and self-pollination may assure the reproductive success, which is common in most of the endemic species. Red sanders is mainly cross-pollinating, indicating that it exhibits facultative xenogamous breeding system, but it also shows geitonogamy (Rao and Raju 2002).

One of the important factors for endemism is its restriction to outcrossing. Pollination biology is less studied in this species. The scope and application of research in the reproduction biology of this species requires a detailed investigation. Ecologists, conservation biologists and breeders need to focus on this grey area and can derive a pragmatic approach to overcome the problems associated with the population expansion of this precious tree. The scattered or little information available is an unknown bottleneck and can draw an immediate attention of the researchers to derive methodologies for the conservation and sustainable utilization of red sanders.

References

Doungyotha Y, Owens JN (2002) The reproductive biology and reproductive success of *Pterocarpus macrocarpus* Kurz. Biotropica 34(1):58–68. https://doi.org/10.1646/0006-3606(2002)034[0058:TRBARS]2.0.CO;2

Endress PK (1995) Diversity and evolutionary biology of tropical flowers. Cambridge University Press, Cambridge

Escobin RP, Payawal PC, Cervancia CR (2004) Pollination syndrome and breeding system of four reforestation tree species in Mt. Makiling, Luzon, Philippines. Philipp Agric Sci 87(2):182–190

Frankie GW, Haber WA, Opler PA, Bawa KS (1983) Characteristics and organization of the large bee pollination system in the Costa Rican dry forest. In: Jones CE, Little RJ (eds) Handbook of experimental pollination biology. Scientific and Academic Editions, New York, pp 411–447

Janzen DH (1967) Synchronization of sexual reproduction of trees within the dry season in Central America. Evolution 21:620–637

Pinheiro M, de Brito VLG, Sazima M (2018) Pollination biology of melittophilous legume tree species in the Atlantic forest in Southeast Brazil. Acta Bot Bras 32(3):410–425

Pal S, Mondal S (2018) Floral biology, breeding system and pollination of *Pterocarpus marsupium* Roxb. Int J Plant Repr Biol 10(2):172–177. https://doi.org/10.14787/ijprb.201810.2

Rao SP, Raju AJS (2002) Pollination ecology of the Red sanders *Pterocarpus santalinus* (Fabaceae), an endemic and endangered tree species. Curr Sci 83:1144–1148

Rao SP, Atluri JB, Reddi CS (2001) Intermittent mass blooming, midnight anthesis and rock bee pollination in *Pterocarpus santalinus* (Fabaceae). Nord J Bot 21:271–276

Propagation of Red Sanders: An Overview

8

M. Anuradha, B. K. Indu, and S. Balasubramanya

Abstract

Pterocarpus santalinus L.f., the pride of Andhra Pradesh, India, is despoiled intentionally many times in want of its wood, which is one of the finest luxury woods of the globe. The commensurate replenishment and sustainable utilization is the only left alternative to preserve this precious species. This chapter emphasizes various natural propagation methods, problems associated with conventional propagation, biotechnological tools to resolve them, and scope of future research.

Keywords

In vitro propagation · Micropropagation · Recalcitrance · Morphogenesis · Regeneration · Acclimatization · Multiple shoots · In vitro rooting

8.1 Introduction

India, 1 of the 12 mega biodiversity countries in the world, abodes many rare, endemic, endangered, and threatened plant species. *Pterocarpus santalinus*, commonly called as red sanders, is one such precious tree which is over-exploited due to multitudinous reasons, and there is a fast depletion of the wild resources and placed in Red List by

M. Anuradha (✉)
Department of Biotechnology, Padmashree Institute of Management and Sciences, Bangalore, Karnataka, India

B. K. Indu
Rishi Foundation, Bangalore, India

S. Balasubramanya (✉)
Department of IT, BT and S&T, Government of Karnataka, Karnataka Innovation & Technology Society, Bangalore, India

T. Pullaiah et al. (eds.), *Red Sanders: Silviculture and Conservation*,
https://doi.org/10.1007/978-981-13-7627-6_8

IUCN (2001). Red sanders is an endemic tree restricted to certain forest tracts of south India. The current disturbance in their narrowly confined ecosystems could mean extermination of the species. Appropriate alternative strategies and replenishment can relieve the exacerbated pressure and can avoid the denudation of natural strands.

The scope for restoration of natural red sanders forests soon is very bleak unless a Herculean effort is taken up for the sustainable development. To ensure sustainable and equitable use of resources for meeting the basic needs of the present and future generations without foreclosing the existing resources, the best option is to practice massive tree plantation of red sanders. Hence attention must be focused to develop a cost-effective and less time-consuming technology for building up elite planting stocks of red sanders and taking up massive plantation drive in all possible geographical locations.

8.2 Plantations and Adaptability

Despite endemic nature of red sanders, the tree is planted and grown to maturity in regions such as Kerala, India (Babu 1992); Sri Lanka (Fernando et al. 2000); Karnataka, India (Arunkumar 2011); and Hyderabad, India (Padmalatha and Prasad 2008). Red sanders planted at the three localities of Kerala has adapted well, and this shows that this endemic taxon can be planted at various other regions and can add to the tree plantation programs, which is linked with afforestation. There is considerable adaptation to varied geographical regions, and this is further established after the successful plantations, raised by Karnataka Forest Department at Kolar, Bangalore, and Mandya (Arunkumar 2011); Nallal of Hoskote Range and Yeshwanthpur area of Kolar Range, Jarakabande, Bangalore; and Hulikere of Mandya forest range, Mandya Division. The status was assessed for variability in growth parameters, heart wood, sap wood, bark thickness, and relationship between girth and heartwood. This versatile nature of the tree directs the environmental scientists and foresters to take up massive plantations. Kumarasinghe et al. (2003) conducted various studies from the seeds collected from the introduced plantations of red sanders in Sri Lanka. The occurrence of fruit-bearing trees indicates its adaptability. This fabulous tree, which formed an irresistible temptation to the smuggler, deserves greater research attention and extensive organized plantings to save it from its current fragile situation. The basic need for massive propagation is establishment of planting stocks in huge quantities. Considering its slow regeneration and growth, prudent management of this limited genotype buttressed with long-term reforestation programs is of cardinal importance and might go a long way in saving the red sanders from depletion (Ahmed and Nayar 1984).

8.3 Natural Regeneration and Conventional Methods of Propagation

Natural regeneration and conventional methods of propagation have been encrusted with various limitations in red sanders. The tree is suitable for growing in tropical regions; however, in natural conditions, when the pods are lying uncovered on the surface of the ground, the radicles dry up during germination, when exposed to the hot sun, and are susceptible to insect damage. Frost is not a problem in the natural habitat, but the seedlings are invariably being killed by frost in the winter (Troup 1921). Ankalaiah et al. (2017) studied population density, structure, and regeneration of red sanders in the protected natural habitat, Sri Lankamalleswara Wildlife Sanctuary, Andhra Pradesh, India. The authors recorded higher percentage of plants in the recruitment stages like saplings and seedlings. Sprouting from roots was observed after fire damage of the existing plants. The bottleneck is its slow growth which is affecting the progression of regenerating sprouts into adult trees. They also assessed population density and suggested that red sanders is tolerant to mild disturbances. But the drastic reduction in the density of the trees with higher girth at breast height class reflects the concern for recruitment in the future as it may affect the seed output due to loss of reproductively fit mature trees. *P. santalinus* does not tolerate overhead shade or waterlogged conditions (Rao and Raju 2002) and is hence seen distributed in the rocky terrains with good water-draining soils. It is slow-growing with prolonged juvenile period and late to reach seed-bearing age. Seed propagation in red sanders is posed with many constraints such as poor pod set, hard fruit coat, and less germinability coupled with poor viability (Dayanand and Lohidas 1988; Naidu and Rajendrudu 2001). Kukrety et al. (2013) reported that seed germination, seedling height, and root collar diameter are significantly stimulated by fire. *P. santalinus*, which is traditionally propagated by seeds, requires a long stratification period to break the dormancy, which partly happens by weathering process of microorganisms (Kalimuthu and Lakshmanan 1995; Arockiasamy et al. 2000). The natural propagation of other *Pterocarpus* species through seeds is also not successful because of poor seed germination as it is also known that the regeneration rate of leguminous trees in natural habitats is quite low (Dewan et al. 1992; Kalimuthu and Lakshmanan 1995).Various trials conducted under field conditions witnessed only 60% germination from seeds collected from Balapally forests of Andhra Pradesh by Kesava Reddy and Srivasuki (1990) and avowed various limitations involved in propagation of red sanders as prolonged dormancy, low germinability, and poor viability of seeds. *P. santalinus* seeds are highly recalcitrant with a dormancy period of 1 year. Poor natural occurrences of germinated seedlings, perhaps due to high temperature, water stress, predation, pod size, light intensity, soil and seed moisture, and seasonal variations, are some of the major problems for its regeneration. Poor germination due to physical nature of the seeds could be appeased to some extent by mechanical or acid scarification (Zodape 1991; Anuradha and Pullaiah 1998). Kumar and Gopal (1975) recorded temperature sensitivity of germination of seeds. Under natural conditions various external factors are also responsible for poor germination apart from the hard fruit coat. As the plant demands strong light, seeds kept in the dark did not germinate;

another reason may be due to nonavailability of nutrient reserves to the newly originated seedling, which is a crucial requirement for the germinated seedling (Anuradha and Pullaiah 1998). Under natural conditions, many seedlings are seen germinating in between the narrow rock crevices, which would have happened due to mechanical stress created due to bulging of the fruit wall after rains. The hard fruit coat and fruit wall leachates were found to inhibit seed germination (Indu and Anuradha, 2018, unpublished data) under in vitro conditions. Fruit walls separated by mechanical means and germinated under in vitro conditions endorsed this resulting in 90–100% germination (Anuradha and Pullaiah 1998; Padmalatha and Prasad 2008). Patel et al. (2018) achieved 66.7% seed germination coupled with improved plant growth and seedling survival rate. Among various methods tried, presoaking of pods with 500 mg/L gibberellic acid for 24 h has relatively better response when compared to tap water, lukewarm water, sulfuric acid, and hydrochloric acid treatments. However, the germination percentage of seeds is highly dependent on geographical location, and there is more germination from seeds procured directly from trees rather than collected from ground. According to Kumarasinghe et al. (2003), only 49% germination was noted even after scarification by alternate drying and soaking when seeds were collected from a plantation at Sri Lanka. They also reported limited fruit-bearing trees located at Sri Lanka. This might be due to endemism and differences in the micro and macro niche conditions. The endemic nature may be due to bigger pod size restricting seed dispersal, phenological constraints, and high vulnerability during early stages of the life cycle. Naidu (2000) carried out scarification of seeds with different concentrations of inorganic acids such as sulfuric, nitric, and hydrochloric acids. Both concentrated and diluted sulfuric acid caused increase in pod germination percentage followed by nitric and hydrochloric acids. Scarification for 50 min with concentrated sulfuric acid resulted in a maximum (81%) germination. Increased duration of scarification reduced the percentage of germination. Naidu (2000) also found that the temperature, in the range of 30–100 °C, was found to enhance germination of red sanders pods subjected to various durations from 5 min to 5 days. The enhancement of germination was found to be dependent on temperature and period of exposure. Pods incubated at 30–100 °C in different durations resulted in 22–84% germination. High percentage (84%) was observed at 90 °C treated for a duration of 20 min and 15 min duration at 100 °C.

Seed size also played a major role to trigger germination, as the rate of germination was higher in larger seeds compared to smaller ones, and this may be attributed to more food reserves which could have stimulated germination, seedling survival, and growth (Milberg and Lamont 1997; Chaturani et al. 2006; Padmalatha and Prasad 2008). The seeds face very heterogeneous situations in field conditions, and other factors such as low viability and decreased fatty acid content due to aging in certain seeds (Thapliyal and Connar 1997) may also contribute for poor germination. This further accelerated loss of moisture, which is associated with a multilayered seed coat, increased conductivity due to leachates and decreased fatty acid content due to aging in certain seeds (Thapliyal and Connor 1997). Chaturani et al. (2006) extensively studied various morphological and maturity parameters responsible for low percentage of germination in red sanders.

The possible reasons for difficulty in conventional propagation of red sanders are not only due to low germination and prolonged dormancy but also because of fungal growth inside the seed coat and seasonal fruit-bearing habit of trees (Kumarasinghe et al. 2003). There were a significant number of publications with respect to improvement of seed germination in red sanders. Naidu (2000, 2001a, b) studied seed pre-treatment methods and the effect of plant growth regulators on seed germination. Among the three plant growth regulators tested, gibberellic acid is found to be more effective in improving germination percentage. Vijayalakshmi and Ranganayaki (2017a) investigated effect of pre-sowing treatment on germination of red sanders. They found that seeds soaked in water for 48 h exhibited better germination response when compared to acid scarification. Vijayalakshmi and Ranganayaki (2017b) performed alternate drying and wetting methods to remove the pod coat. They pronounced that out of 96% viable seeds, 73% germinated and 23% neither germinated nor dead even after imbibitions. Patel et al. (2018) also performed experiments on germination of red sanders and concluded that 500 ppm GA_3 treatment is the best in terms of sprouting percentage, collar diameter, plant height, number of leaves per plant, survival percentage, root length, and fresh and dry weight of the plant. This clearly indicates the fact that the less percent of germination may also be due to seed dormancy factors and hard pod coat. Sankanur and Shivanna (2010) in their attempt to improve growth in *P. santalinus* seedlings at field level tried integrated nutrient management. In the process they found that application of 33 g of poultry manure showed maximum seedling height, collar diameter, number of leaves, and total fresh and dry weight.

Forestry researchers are encouraged to undertake the comprehensive trials necessary for establishing commercial plantations. Among them, top priority should be given to acquire superior varieties (wavy grained, ripple wood, and high santalin), seed production, storage and testing the seed for its purity of producing parental characters, germplasm storage, and performance at different ecological zones. Vegetative propagation is one of the best means when it comes to multiplication of selected trees, especially trees like red sanders as there is a mixed population of elite and non-elite trees.

Clonal multiplication through rooted cuttings is also an obstinate task in red sanders. When compared to seed germination experiments, there are almost no studies on ex vitro vegetative propagation in red sanders (Teixeira et al. 2018). Executed with reported conventional vegetative propagation methods such as semi-hard wood cuttings, cleft grafting, and air layering were executed with varying degrees of success and has limitations in producing enough numbers of planting material for forestry programs (Kesava Reddy and Srivasuki 1990). Kedharnath et al. (1976) tried to unite two variants of red sanders and recorded 21.74% survival of the grafts. The snags encountered for large scale multiplication of this species by conventional vegetative propagation methods such as grafting, and air-layering is ascribed to its poor performance in rooting the cuttings (Kesava Reddy and Srivasuki 1990). Some degree of success has been attained by modification of air layering treatments by Kesava Reddy and Srivasuki (1992) where they employed 1000 ppm of indole-3-butyric acid. This process induced roots in about 3 months, and rooting was preceded with callus formation in all the cases. In continuation they tried rooting the coppice cuttings with success rate of 50%. Due to this, multiplication of elite trees on a large scale is challenging, forcing them to identify alternative methods for clonal propagation. emphasis on constraints of natural regeneration methods via seeds or rooted cuttings, as this offer cheaper and easy to scale up. "with" an understanding about natural regeneragtion medthods which are cheaper and easier to scale up.

After enough literature survey, the inference is red sanders is not amenable to conventional methods of propagation, though preferred for building up planting stocks. Prolonged juvenile period of this tree takes many years to reach seed-bearing age. On the other hand, seed propagation also has major constraints such as prolonged dormancy, low germinability, differential seed size accounting for immaturity, phenolics in the seed coat, hard fruit coat, vulnerability during early seedling stage, grazing, forest fires, and poor viability. Clonal multiplication through rooted cuttings has been attempted for many years and discussed in this chapter. Vegetative propagation also forms a hindrance as it cannot be applied to red sanders; if so practiced, the number of cuttings and size of the cuttings are more which in turn question the very existence of the elite mother plant. Further, the cutting can be procured only during a specific season, not round the year. When such conventional methods of propagation achieve poor and impracticable results, it has become imperative to devise methods by which large-scale propagules of selected elite can be raised.

Hence there should be deployment of modern technologies like micropropagation for building up planting stocks in red sanders. potentially by providing tool a mammoth potential to produce massive planting stocks, true to type plants and preservation of germplasm.

8.4 Need for Tissue Culture Studies

With this perspective, the technique of plant tissue culture has precisely been envisaged as an effective means with colossal potentiality, offering methods for rapid multiplication to build substantial number of planting stocks for afforestation programs and conservation of rare and elite germplasm of red sanders. This is unerringly proven beyond doubt, as the technology untangled many problems encountered in forestry and conservation of tree species. The application of in vitro technique has been well established in many crop plants and ornamentals particularly in herbaceous species. However, progress in establishing in vitro cultures of woody plants has been limited to a handful of taxa. Plant cell tissue and organ culture now provides an array of techniques to produce viral-free plants through meristem culture, anther culture to establish homozygous lines and most significantly micropropagation for obtaining true-to-type plants. However, adaptation of this technology to betterment of the leguminous trees through tissue culture is not done in the same pace along with other herbaceous species, as regeneration of forest trees particularly legumes has been an arduous task. Considering red sanders as a deserving species for conservation and sustainable utilization, the amount of work done to propagate via tissue culture methods is limited due to its recalcitrant nature.

8.5 Tissue Culture Studies

The precedent set in the past asserts red sanders as one of the tough materials for tissue culture studies. Forest trees in general and legumes are recalcitrant to regeneration (Sita et al. 1992). With the advancement in plant tissue culture research, there is a tremendous progress, and various reports have appeared on regeneration of woody plants. However, there are several unexploited forest trees, including red sanders, which need to be studied in vitro. There is a dire need to optimize technology for selected clones of elite varieties of red sanders. A rapid propagation technique is highly desirable as the traditional in vivo methods of propagation are too slow or unprofitable as there is a limitation of rooting in turn affecting the number of stock plants produced. The endemic nature of the tree and nonavailability of starting material in vicinity coupled with problems associated with engendering tissue cultures from the mature tree explants would have prompted many researchers to deploy seeds or seedlings as the starting material. Majority of publications dealt with seedlings or seed cultures and the attempts of micropropagation in red sanders in limited to handful of publications, leaving incredible scope for future research.

8.6 In Vitro Seed Cultures

Seed culture is the first and foremost option for plant tissue culturist to establish clean mother cultures in vitro, and this is undeniably an excellent option in case of tree tissue culture. This can solely be attributed to recalcitrant nature of mature tree explants when used as the starting material. The drawbacks of conventional seed germination and vegetative propagation in *Pterocarpus santalinus* are unswervingly deliberated by many researchers (Ahmed and Nayar 1984; Dayanand and Lohidas 1988; Kesava Reddy and Srivasuki 1990; Kalimuthu and Lakshmanan 1995). Similar constraints would be faced while establishing in vitro seed cultures. Under innate situation, red sanders propagation through seeds is curtailed by poor pod set (Troup 1921), presence of phenolic compounds in seed coat (Venkataramaiah et al. 1980), and low percent of germination (Ahmed and Nayar 1984). Though the percentage of germination is more in seed sowing, the possibility of seed damage while separating the hard fruit coat was reported by Dayanand and Lohidas (1988). Hence before processing for in vitro seed cultures, fruits need to be subjected to various treatments for easy pod removal to avoid mechanical damage. Anuradha and Pullaiah (1998) made successful attempts and optimized a protocol for easy fruit coat removal. The objective was achieved by soaking the fruits in acid alcohol mixture (1:1 volume/volume) for 2 h and washed under running tap water to remove the traces of acid alcohol. This resulted in 100% damage-free seeds and further employed in vitro seed cultures. Under in vitro conditions, though seeds exhibited better germination, the establishment of seedlings into healthy plants is in question, and half strength B5 medium, fortified with small amounts of BAP, was found to be more responsive. Chaturani et al. (2006) performed in vitro seed germination studies and correlated germination percentage with morphology of pod, size, seed color, harvest time, and storage period. Seeds cultured within 1 week of harvest exhibited about 96% of in vitro germination. This is also highly varied with type of media used, and according to them germination percentage is significantly higher in Anderson medium without charcoal and relatively low percentage of germination on Woody plant, MS, and Vitis medium. Vipranarayana et al. (2012) stressed the importance of GA_3 in the media for successful seed germination. Though in vitro germination is achieved with different patterns by various researchers, the major problem encountered is establishment of healthy in vitro cultures as there is predominant shoot-tip necrosis in cultures leading to mortality (Anuradha and Pullaiah 1999a, b).

8.7 In Vitro Propagation

The importance of the red sanders and reasons for conservation have evinced interest in many researchers and resulted in attempts to optimize of suitable protocols for in vitro propagation. To establish an efficacious large-scale propagation, the basic requirements are establishment of primary explants in culture and optimizing the supplements of culture media. The initiation of callus induction, organogenesis, and growth and development of regenerated organs are considered as separate events with different requirements. During the process of in vitro development callus may not always differentiate or the differentiated organs may not develop further. Hence specific attention must be focused for standardizing each step. In any tree species when the objective is to attain true-to-type plants, it is always desirable to focus on callus-free micropropagation system, and this is also applicable to red sanders.

8.7.1 Initiation and Establishment of Mother Cultures

Any effective tissue culture protocol highly relies upon the type of explant and optimization of surface sterilization methods. In red sanders, the initial explant is sourced either from field-grown seedlings (Balaraju et al. 2011) or from aseptically grown seedlings. Patri et al. (1988) treated the seeds with 40% HCl for 24 h. Before excising they were rinsed in 90% ethanol and surface sterilized with 15% sodium hypochlorite solution for 15 min and washed with sterile distilled water. However, only 17% seed germination was reported on MS media. Seeds can withstand harsh surface sterilization methods, as they are sufficiently hardy; however, the difficulty is with delicate explants. Sita et al. (1992) used shoot tips excised from 15-day-old seedlings raised in seed beds. Detergent treatment followed by 0.1% mercuric chloride treatment is adapted to initiate tissue cultures. The usage of mercuric chloride as a surface sterilant is predominant in red sanders (Anuradha and Pullaiah 1999a, b; Balaraju et al. 2011; Sita et al. 1992; Padmalatha and Prasad 2008). This is the most critical factor which decides the fate of explants in micropropagation and not conferred due importance and very limited studies were made in this regard.

Particularly when the coppices of elite material are procured after surpassing many difficulties, explant is very precious and need to be rescued from microbial contamination. In red sanders microbial contamination and phenolics are the major constraints in establishing tissue cultures from seeds, particularly from mature tree explants. Hence it is very much imperative to optimize the surface sterilization methods along with removal of phenolics from the explants for a successful method standardization. Chaturani et al. (2006) asserted the importance of surface sterilization in establishing clean cultures of red sanders.

The choice of the explant is another vital factor while developing a process for micropropagation and in tree tissue culture screening the explants for successful establishment of mother cultures is a challenging task. The general recalcitrance nature of leguminous species for morphogenesis has been highlighted (Mohan Ram et al. 1981; Sita et al. 1992). The failures reported by many researchers may be partly due to the paucity of assessment of explants for morphogenic potentiality. The first report on micropropagation of *Pterocarpus santalinus* reveals the usage of cotyledon explants from aseptically raised seeds. Though multiplication rate is not encouraging, axillary branches were excised and subcultured for further multiplication, and this is accounted for fivefold multiplication rate within 35 days (Patri et al. 1988). A comprehensive list of publications deploying different explants is reviewed (Table 8.1).

Table 8.1 Response of different explants cultured in vitro in *P. santalinus*

Source	Explant	Response	Author
In vitro raised seedlings (30 days old)	Nodes and terminal cuttings	2 shoots	Patri et al. (1988)
Mature trees (15 years old)	Shoot tips	Bud break	Kesava Reddy and Srivasuki (1992)
Seedlings from seed bed (15 days old)	Shoot tips, nodes	Bud break and 2–3 shoots	Sita et al. (1992)
In vitro seedlings (15 days old)	Mesocotyl	5–8 shoots	Anuradha and Pullaiah (1999b)
In vitro germinated seedlings (21 days old)	Cotyledon	8–10 shoots	Arockiasamy et al. (2000)
Mature trees	Seeds	3 shoots	Chaturani et al. (2006)
10-year-old trees	Nodal explants	5–6 shoots	Prakash et al. (2006)
Seedlings from seed bed (30 days old)	Cotyledonary nodes	3–4 shoots	Rajeswari and Paliwal (2008)
In vitro seedling (35 days old)	Cotyledonary nodes and meristem	10–12 shoots	Padmalatha and Prasad (2008)
Seedlings from field (20-day seedling)	Shoot tips	6 shoots	Balaraju et al. (2011)
Germinated seeds (21 days old)	Nodal segments	11 shoots	Vipranarayana et al. (2012)
In vitro seedlings (20 days old)	Cotyledonary nodes	5 shoots	Warakagoda and Subasinghe (2013)

Anuradha and Pullaiah (1999b) cultured root, hypocotyl, mesocotyl, cotyledon, shoot tip, nodal, leaf, and internodal explants and reported differential morphogenetic responses. Mesocotyl, shoot tip, and nodal explants resulted in shoot bud regeneration, whereas other explants callused from the cut edges. This is further confirmed by Arockiasamy et al. (2000) who reported that only shoot tips and axillary nodes can regenerate multiple shoots and there is no shoot formation from epicotyl, hypocotyl, and internode explants. Padmalatha and Prasad (2008) employed seedling explants and successfully established in vitro cultures and reported four to six shoots. The choice of explant for success in micropropagation in red sanders is immutable, and this was studied by Warakagoda and Subasinghe (2013).

In vitro clonal multiplication of *Pterocarpus santalinus* was achieved by Prakash et al. (2006) using mature nodal explants of a 10-year-old elite quality tree. Combinations of serial transfer technique and incorporation of antioxidants (250 mg/l L-ascorbic acid and 50 mg/l citric acid) into the culture medium helped to minimize medium browning and improve explant survival during shoot sprouting. The explant harvest period also influenced the bud break and shoot sprouting in nodal explants.

It is an engrained fact that media constituents also have pivotal role in morphogenesis and shoot multiplication. Media and growth regulators are two major components which will govern the response in plant tissue cultures. There are limited attempts to screen media for inducing positive response in red sanders. But the type of media invariably influenced the morphological nature of the shoots regenerated. Anuradha and Pullaiah (1999b) optimized and derived an efficient system for regenerating healthy multiple shoots of red sanders. The explants were cultured on B5 medium initially and later were transferred to MS medium for healthy growth without precocious leaf drop and necrosis. This was further improved by reducing the subculture period to 10 days. Padmalatha and Prasad (2008) experimented with MS, McCown, and Gamborg media and did not find any differences in the response and hence continued with MS medium. Prakash et al. (2006) reported that about 70% of explants were sprouted on Murashige and Skoog (MS) liquid medium containing 4.4 μM 6-benzyladenine (BA).

The undisputable role of growth regulators and media adjuvants in morphogenesis and shoot multiplication was well demonstrated in *P. santalinus*. Patri et al. (1988) augmented percentage of germination with two shoots each, when seeds were inoculated on BAP fortified media. The usage of adenine and glutamine also enhanced shoot multiplication and the rate declined later. This was attributed to the endogenous accumulation of certain factors which inherently responsible for the reduced response. Leaf drop is a problem in red sanders tissue culture was sorted out by addition of adenine. Necrosis was reported by Patri et al. (1988), however unraveled by initiating cultures on B5 medium and transferring to MS media. This is further improved by adopting quick passages by Anuradha and Pullaiah (1999b). Adenine and guanine fortification to the media promoted shoot multiplication, but the rate of shoot proliferation declined after subsequent passages and attributed to the accumulation of endogenous substances (Patri et al. 1988).

A detailed study and relatively better response in terms of multiple shoot proliferation is narrated by Sita et al. (1992) in red sanders. The authors envisaged that the supplementation of the combination of benzyladenine and kinetin benefited the individual cytokinin fortification and enhanced the multiplication of shoot buds. Eight shoot buds were regenerated from shoot tips cultured on B5 media fortified with benzyladenine and kinetin within 4–6 weeks. Patri et al. (1988) reported regeneration of shoots with scaly leaves when single cytokinin is used, whereas this problem is resolved by employing combination of cytokinins by Sita et al. (1992). Prakash et al. (2006) reported that the combination of 4.4 μM BA and 2.2 μM thidiazuron (TDZ) was found to be the most suitable growth regulator for obtaining the highest percentage of nodal segment sprouting (74–75%), the number of secondary shoots per primary shoot (two or three), the shoot length (5–6 cm), the number of new nodal segments generated per active explant (four or five), and the multiplication coefficient (3.5) within 6 weeks. Repeated subculturing of nodal explants obtained from shoot cultures enabled continuous production of healthy axillary shoots. At the end of the sixth passage, about 90% of nodal explants produced five or six healthy green shoots, each being about 6.6 cm long with six or seven nodes. Multiplication coefficient was also increased from the first subculture (5.4) to the sixth subculture (8.3).

Padmalatha and Prasad (2008) studied the effect of TDZ in media on shoot multiplication, which were found to induce a maximum of four shoots. TDZ when fortified in combination with BAP resulted in six to seven shoots coupled with hard brownish callus. Rajeshwari and Paliwal (2008) emphasized the significant role of growth regulators in morphogenesis; nevertheless the choice of explant was also a notable factor in determining the success of micropropagation of red sanders. Balaraju et al. (2011), in their study, reiterated that the combination of BAP and TDZ is most effective both to induce bud break and multiple shoot regeneration, and maximum of 11 shoot buds were noticed. In contrary Vipranarayana et al. (2012) expressed that BAP was superior to TDZ and induced relatively more number of shoot buds. However, Anuradha and Pullaiah (1999b) reported 10–15 shoots from mesocotyl explants cultured on B5 medium supplemented with 2 mg/l KN and 1 mg/l BAP.

8.7.2 Recalcitrant Nature of Red Sanders

Culturing mature tree explants needs to be preferred over seedling explants because it is often not possible to determine whether these embryos or seedlings have the genetic potential to develop the desired qualities and because of their heterozygous nature. However, most of the studies carried in red sanders have utilized seeds and seedling explants that are more amenable to in vitro manipulations than tissues for mature plants. The reasons as discussed earlier may be accumulation of phenolics, high degree of contamination, physiological variations due to seasonality, and loss of morphogenetic potentiality. The success of establishing tissue cultures from mature tree explants in red sanders is limited, and there were sporadic reports regarding this. As said earlier bud break from 10-year-old mature tree explants was reported by Prakash et al. (2006). They are successful in reducing browning adopting serial transfer technique coupled with usage of antioxidants. Supplementation of dual cytokinins, BA, and Thidiazuron exhibited positive influence with respect to red sanders. Best rooting response resulted in IBA-supplemented media, and 70% of micropropagated plants were successfully acclimatized.

8.7.3 Rooting

In vitro or ex vitro rooting is one of the important stages which will help the technology transfer from lab to land. Many of the recalcitrant species, including red sanders, are difficult to root even under in vitro conditions. The pioneering report of micropropagation of *P. santalinus* did not endeavor any work on rooting of in vitro regenerated shoots (Patri et al. 1988). Sita et al. (1992) are the first team of researchers to reveal successful rooting by employing IAA, NAA, and IBA. Among the three auxins they tried, IAA was found to exhibit relatively better response with 80% of rhizogenic potentiality. Microshoots regenerated were initially treated with various concentrations of auxins, and then transferring to basal media is effective to induce roots (Anuradha and Pullaiah 1999b). Arockiasamy et al. (2000) induced roots by implanting excised shoots on one-fourth strength MS medium supplemented with different concentrations of IAA and IBA and testified nil effect of NAA on in vitro rhizogenesis. In similar lines the positive effect of IBA on root regeneration was mentioned by Balaraju et al. (2011). In contrast Padmalatha and Prasad (2008) experimented with all the three auxins IAA, NAA, and IBA, which were found to be ineffective to induce rooting. When they employed MS basal media and rooted the plantlets, however efficient rooting was recorded when transferred the shoots in bunches of three to four. Warakagoda and Subasinghe (2013) achieved 40% rooting by pulse treatment with auxins.

8.7.4 Acclimatization and Hardening

Though many publications ponder on micropropagation of red sanders, very less attention is being paid toward this decisive stage. Sita et al. (1992) successfully established 50% of micropropagated plantlets in sterile soil mixtures by following gradual acclimatization process. Maintaining high humidity conditions after transferring to soil and sand mixture is necessary for hardening red sanders (Anuradha and Pullaiah 1999b). Ninety percent survival was reported by Arockiasamy et al. (2000) by transferring plantlets to paper cups containing a mixture of sterilized soil and sand. The plants were gradually hardened by irrigating with one-fourth strength MS medium fortified with 1 mg/l IAA followed by tap water for 1 week. For the first 15 days, these plantlets were maintained under controlled temperature and light conditions and then transferred to greenhouse conditions. These hardened plants eventually transferred to field conditions and provided Hoagland solution for

1 week. Balaraju et al. (2011) screened different hardening media and reported 73% survival rate on organic manure and sand mixture, which was found to be the highest. After 45 days of hardening, plants were transferred to pots containing soil. Padmalatha and Prasad (2008) used soil rite media for acclimatization and reported 20% survival rate. Warakagoda and Subasinghe (2013) reported 80% success rate while hardening red sanders in vitro regenerated plants by following the two-stage acclimatization process.

Rooting in vitro regenerated plants by ex vitro methods is an excellent concept, which has multiple benefits. Directly transferring plants to soil media will reduce time and has an advantage of simultaneous hardening and rooting. This process also reduces cost of production of plants by micropropagation. Many times, though rhizogenesis is induced in vitro, new roots regenerated after transferring to the soil conditions ensuring for better field performance. This technique is followed by Rajeswari and Paliwal (2008), to acclimatize red sanders by screening various hardening media. Among all, a mixture of coarse sand, clay, and farmyard manure mixed in equal proportions resulted in maximum percentage of survival. Vipranarayana et al. (2012) performed similar experiments and reported 85% of rooting response and claimed the ex vitro rooting method to be promising and attractive, which reduces the cost and time required for establishment in the field.

Field performance study of in vitro raised plants is not done with the same pace as in vitro multiplication study. Any successful micropropagation study should conclude with best lab to land performance. Unfortunately, very little importance is given for this and hence restricted utility of the technology in red sanders. Rajeswari and Paliwal (2008) evaluated growth patterns of tissue-cultured plants in comparison with conventionally propagated plants. The results indicated that the tissue-cultured plants exhibited better performance and could be adopted for large-scale in vitro propagation of red sanders.

8.7.5 Scope and Prospects

The conservation and sustainable utility of red sanders is deliberated widely by all segments of the society. There are many unanswered questions such as identification of superior trees at the nursery stage, scaling up of optimized micropropagation methods, entangling the problems associated with establishment of tissue cultures from mature tree explants, development of reproducible lab to land technologies, feasibility studies, marker-assisted selection, and large-scale multiplication of elite varieties. Recent waning in plant tissue culture research, due to divergence to many advanced focus areas, would have led to the decline in research and development of red sanders. However, there is a warning signal for all the plant biologists to take up this study, and conglomerate efforts are a must to protect and conserve this medicinal and luxury wood-yielding timber tree.

Acknowledgments First author Indu BK thanks the Department of Science and Technology, New Delhi, for the financial support under DST (WOS-A) scheme.

References

Ahmed M, Nayar MP (1984) Red Sanders tree (*Pterocarpus santalinus* L.f.) on the verge of depletion. Bull Bot Surv India 26:142–143

Ankalaiah C, Mastan T, Reddy MS (2017) A study on the density, population structure and regeneration of red sanders *Pterocarpus santalinus* (Fabales: Fabaceae) in a protected natural habitat- Sri Lankamalleswara Wildlife Sanctuary, Andhra Pradhes, India. J Threat Taxa 9 (9):10669–10674

Anuradha M, Pullaiah T (1998) Investigations on germination of enhanced axillary branching in *Pterocarpus santalinus* (Red sanders) with special reference to in vitro seed culture. Indian For 124:309–314

Anuradha M, Pullaiah T (1999a) In vitro seed culture and induction of enhanced axillary branching in *Pterocarpus santalinus* and a method for rapid multiplication. Phytomorphology 49:157–163

Anuradha M, Pullaiah T (1999b) Propagation studies of Red sanders (*Pterocarpus santalinus* L.f.) in vitro- an endangered taxon of Andhra Pradesh, India. Taiwania 44(3):311–324

Arockiasamy S, Ignacimuthu S, Melchias G (2000) Influences of growth regulators and explants type on in vitro shoot propagation and rooting of red sandal wood (*Pterocarpus santalinus* L.f.). J Exp Biol 48:1270–1273

Arunkumar AN (2011) Variability studies in *Pterocarpus santalinus* in different aged plantations of Karnataka. My For 47(4):343–353

Babu NVT (1992) Introduction of Red sanders in Kerala. Indian For 118(2):109–111

Balaraju K, Agastian P, Ignacimuthu S (2011) A rapid in vitro propagation of red sanders (*Pterocarpus santalinus* L.) using shoot tip explants. Acta Physiol Plant 33:2501–2510

Chaturani GDG, Shubhashinge S, Jayatillake MP (2006) In vitro establishment, germination and growth performance of Red sandal wood (*Pterocarpus santalinus* L.). Trop Agric Res Ext 9:116–130

Dayanand T, Lohidas T (1988) Studies on development and maturity of Red sanders (*Pterocarpus santalinus* Linn. f.). Indian J For 11:207–208

Dewan A, Nanda K, Gupta SC (1992) In vitro micropropagation of *Acacia nilotica* subsp. *indica* Brenan via cotyledonary nodes. Plant Cell Rep 12:18–21

Fernando KCE, Dayatilake GA, Senaratne R (2000) Studies on the propagation of Red sandal wood (*Pterocarpus santalinus* Linn. f.), an endangered medicinal tree species. In: Proc. 6th annual forestry and environment symposium. University of Sri Jayewardenepura, Sri Lanka, p 44

Indu BK, Anuradha M (2018) Optimization of marker assisted technology for screening, selection and micropropagation of elite genotypes of *Pterocarpus santalinus* (Red Sandalwood). Project report submitted to Dept. Sci. Tech., Government of India (unpublished). http://online-wosa.gov.in:8080/wosa/wos/projectReceived_viewProjectRecieved.action. (2014). p 75

IUCN (2001) IUCN red list of threatened plants, species survival commission. IUCN, Gland

Kalimuthu K, Lakshmanan KK (1995) Effect of different treatments on Pod germination of *Pterocarpus* species. Indian J For 18:104

Kedharnath S, Rawat MS, Uniyal DP, Kantham DL (1976) Studies on field grafting and the growth of grafts in Red Sanders. Indian For 102:761–765

Kesava Reddy K, Srivasuki KP (1990) Vegetative propagation of Red Sanders (*Pterocarpus santalinus* Linn.f.). Indian For 116(7):536–540

Kesava Reddy K, Srivasuki KP (1992) Biotechnological approach for tree improvement in Red sanders. Indian For 118:1–4

Kukrety S, Gezan S, Jose S, Alavalapat JRR (2013) Facilitating establishment of advance regeneration of *Pterocarpus santalinus* L. an endangered tree species from India. Restor Ecol 21 (3):372–379

Kumar A, Gopal M (1975) A note on temperature sensitivity for germination of *Pterocarpus santalinus* seed. Seed Sci Technol 3:605–613

Kumarasinghe HKMS, Subasinghe S, Aruna Kumara KKIU (2003) Seed propagation of Rath Handun (*Pterocarpus santalinus* L.) as affected by method of soaking and scarification. In: Proc. 9th annual forestry and environment symposium. University of Sri Jaye-wardenepura, Sri Lanka, p 57

Milberg P, Lamont BB (1997) Seed/cotyledon size and nutrient content play a major role in early performance of species on nutrient-poor soil. New Phytol 137:665–672

Mohan Ram HY, Mehta U, Rao IVR (1981) Tissue and protoplast culture and plantlet regeneration in legumes. In: Rao AN (ed) Proceedings of the costed symposium on tissue culture of economically important plants, Singapore

Naidu CV (2000) Seed scarification requirement in *Pterocarpus santalinus* Linn. f. J Indian Bot Soc 79:175–178

Naidu CV (2001a) Improvement of seed germination in Red sanders (*Pterocarpus santalinus* Linn. f.) by plant growth regulators. Indian J Plant Physiol 6:205–207

Naidu CV (2001b) Seed pretreatment methods to improve germination in red sanders (*Pterocarpus santalinus* Linn. f.). Indian J For 24:342–343

Naidu CV, Rajendrudu G (2001) Influence of kinetin and nitrogenous salts on seed germination of red sanders (*Pterocarpus santalinus* Linn. f.). Seed Sci Technol 29:669–672

Padmalatha K, Prasad MNV (2008) In vitro plant regeneration of *Pterocarpus santalinus* L.f (Red sanders) – an endangered medicinal plant and important timber tree. Tree For Sci Biotechnol 2 (1):1–6

Patel HS, Tandel MB, Prajapati VM, Amlani MH, Prajapati DH (2018) Effect of different treatments on germination of red sanders (*Pterocarpus santalinus* L.f.) in net house condition. Int J Chem Stud 6:876–879

Patri S, Bhatnagar SP, Bhojwani SS (1988) Preliminary investigation of a leguminous timber tree: *Pterocarpus*. Phytomorhology 38(1):41–45

Prakash E, Shavalikhan PS, Srinivasa Rao TJV, Meru ES (2006) Micropropagation of Red sanders (*Pterocarpus santalinus* L.f.) using nodal explants. J For Res 11(5):329–335

Rajeswari V, Paliwal K (2008) In vitro plant regeneration of red sanders (*Pterocarpus santalinus* L. f.) from cotyledonary nodes. Indian J Biotechnol 7:541–546

Rao SP, Raju AJS (2002) Pollination ecology of the Red Sanders *Pterocarpus santalinus* (Fabaceae), an endemic and endangered tree species. Curr Sci 83(9):1144–1148

Sankanur MS, Shivanna H (2010) Effect of integrated nutrient management on growth and development of the *Pterocarpus santalinus* Linn.f. seedlings. Karnataka J Agric Sci 23 (5):726–728

Sita GL, Sreenatha KS, Sujata S (1992) Plantlet production from shoot tip cultures of red sandal wood (*Pterocarpus santalinus* L.f.). Curr Sci 62(7):532–535

Teixeira JA, Kher MM, Soner D, Nataraj M (2018) Red sandalwood (*Pterocarpus santalinus* L.f.): biology, importance, propagation and micropropagation. J For Res. https://doi.org/10.1007/s11676-018-0714-6

Thapliyal RC, Connor KF (1997) Effects of accelerated ageing on viability, leachate exudation and fatty acid content of *Dalbergia sissoo* Roxb. seed. Seeds Technol 25:311–319

Troup RS (1921) *Pterocarpus*. In: The silviculture of Indian trees, vol 1. Clarendon Press, Oxford, pp 265–227

Venkataramaiah V, Prasad SV, Swamy RR (1980) Levels of phenolic acids in (*Pterocarpus santalinus* L.f.). Indian J Exp Biol 18:887–889

Vijayalakshmi KP, Renganayaki PR (2017a) Effect of pre-sowing treatment on germination of Red sanders. Int J Curr Microbiol App Sci 4:168–173

Vijayalakshmi KP, Renganayaki PR (2017b) Seed germination improvement studies in Red sanders (*Pterocarpus santalinus* L.f.) from Karnataka. Int Res J Multidiscip Sci Technol 2(1):1–5
Vipranarayana S, Prasad TNKV, Damodharam T (2012) In vitro seed germination and induction of enhanced shoot multiplication in *Pterocarpus santalinus* Linn. f: an endemic medicinal plant of Seshachalam hills, Tirumala. Int J Pure Appl Sci Technol 9(2):118–126
Warakagoda P, Subasinghe S (2013) In vitro propagation of *Pterocarpus santalinus* L. (Red Sandalwood) through tissue culture. J Natl Sci Found Sri Lanka 41(1):53–63
Zodape ST (1991) The improvement of germination of some forest species by acid scarification. Indian For 3:61–66

Silviculture of *Pterocarpus santalinus* 9

B. N. Divakara and C. U. Nikhitha

Abstract

This chapter deals with silviculture of red sanders, one of the finest luxury timbers in the globe. *P. santalinus* is limited to Southern tropical dry deciduous forests at altitude ranging from 150 to 900 m. Details of geoclimatic factors like climatic conditions, topography, geology, and soil where the tree is naturally distributed are mentioned. Phenological details, natural regeneration, and vegetative methods of propagation are discussed. Silvicultural practices, diseases, and population studies are highlighted.

Keywords

Pterocarpus santalinus · Phenology · Distibution · Population status · Silvicultural practices

9.1 Introduction

Pterocarpus is a genus of trees belonging to family Fabaceae, classified under Dalbergieae tribe. Most of the species of *Pterocarpus* group yield valuable hardwood widely used as timber. They have gained high importance for their ethnomedicinal uses and valuable bioactive compounds (Kumar and Joshi 2014). The most important *Pterocarpus* species are *Pterocarpus santalinus*, *P. cambodianus*, *P. dalbergioides*, *P. indicus*, *P. marsupium*, *P. macrocarpus*, and *P. pedatus* widely distributed across the world especially India, China, Sri Lanka, and Africa (Wenbin and Xiufang 2013). Of these, only four species are found in India, *Pterocarpus santalinus*, *P. dalbergioides*, *P. indicus*, and

B. N. Divakara (✉) · C. U. Nikhitha
Institute of Wood Science and Technology, ICFRE, Bangalore, Karnataka, India

T. Pullaiah et al. (eds.), *Red Sanders: Silviculture and Conservation*,
https://doi.org/10.1007/978-981-13-7627-6_9

P. marsupium. Among the four species, *P. santalinus* is classified as endangered, and *P. marsupium* and *P. indicus* are listed as vulnerable by the International Union for Conservation of Nature (IUCN) (Prasad et al. 2008; Lakshminarasimhan and Mondal 2012).

P. santalinus, also known as "pride of Eastern Ghats," is resistant to drought, which falls under Deccan thorn scrub and Central Deccan dry deciduous forests eco-terrestrial region (GOI 2014). The plants grow well in lateritic and gravel soil (GOI 2014), as they avoid waterlogging and overshading by other trees (Raju and Raju 2000). In international market *P. santalinus* is highly valued for its heavy, dark claret-red heartwood having array of uses in furniture, making carvings, musical instruments, medicine, and extraction of dye (Bhagyaraj 2017). The trees of red sanders are strongly allied with *Pterocarpus marsupium*, *Hardwickia binata*, *Anogeissus latifolia*, *Chloroxylon swietenia*, and *Albizia lebbeck* (GOI 2014). However, many times they are also found in pure patches (GOI 2014).

Red sanders being a valuable timber with various uses (Ankalaiah et al. 2017) is an extensive tradable commodity since the sixteenth century (Vedavathy 2004; Kumar and Joshi 2014). "Santalin," a phytochemical extracted from heartwood, is used as coloring agent in pharmaceutical preparations, food items, leather, and textile industries (Vedavathy 2004; Pandey et al. 2014). The paste/tonic prepared from its wood (Ankalaiah et al. 2017) is used as traditional medicine for multiple uses (Azamthulla et al. 2015). The timber is in extensive demand in Japan and China for its exquisite color and superior acoustic qualities in preparing musical instruments, luxury furniture, and carvings (Ankalaiah et al. 2017; Azamthulla et al. 2015).

The heavy demand in the international market and continuous pressure in harvesting the trees extensively have led to overexploitation in their natural habitat through large-scale smuggling (Kumar and Joshi 2014; Mahammadh 2014). In present days, *P. santalinus* is gaining importance on farmers' field because of its highly valued timber. However, encouraging cultivation on farmlands is hindered because of lack of knowledge on its silviculture requirements. Thus, this chapter focuses on silviculture of *P. santalinus*.

9.2 Distribution and Forest Type

Pterocarpus santalinus, popularly known as red sanders, almug, and saunderswood, is a typical leguminous species endemic to the tropical dry deciduous forests of the southern Eastern Ghats (Ankalaiah et al. 2017; Ahmed and Nayar 1984; Jadhav et al. 2001). Significantly these trees are confined to forests of Andhra Pradesh with natural hilly terrains and dry slope altitude habitat (Ankalaiah et al. 2017). They are scattered across the parts of Chittoor and Kadapa districts and extending slightly to Kurnool, Prakasam, and Nellore districts of Palakonda and Seshachalam hill ranges and distributed over an approximate area of 5160 km^2 (GOI 2014). They are also scarcely found in the regions of Vellore (former North Arcot) and Chengalpattu districts of Tamil Nadu and Karnataka. Their occurrences were reported in other countries like China, Pakistan, Sri Lanka, the Philippines, and Taiwan, ought to be an introduced species (Kumar and Sane 2003; Mulliken and Crofton 2008).

P. santalinus is restricted to the southern east portion of the Indian Peninsula. Predominantly, they are found in Kadapa district of Andhra Pradesh, but the small portions are grown in the adjacent districts of Chittoor, Kurnool, and Nellore in Andhra Pradesh and in Vellore (North Arcot) and in the outskirts of Kambakkam and Nagalapuram hills of Chingleput district in Tamil Nadu basically in the Seshachalam, Palakonda, Lankamala, and Veligonda Ranges of hills. The tree species like red sanders and *P. dalbergioides* are very significant in India with limited distribution (Troup 1983).

According to Andhra Pradesh Forest Development Corporation, it approximately grows in 50, 000 ha of forest area. The natural range is restricted to typically dry, hilly, often rocky ground, at altitudes of 150–900 m, in areas receiving around 100 mm of rain in each of the two annual monsoons. The tree does not tolerate overhead shade or waterlogged conditions (Kumar 2011).

Red sanders trees are reported in the eight forest divisions of Andhra Pradesh, with range varying from 200,000 to 219,000 ha and to 398,000 ha. In comparison with the eight forest divisions of Andhra Pradesh, Kadapa (105,000 ha), Rajampet (85,000 ha), and Proddatur (78,000 ha) are reported as largest species bearing forests. But naturally, trees are distributed in approximately 4 lakh hectares of Seshachalam, Veligonda, Lankamala, and Palakonda hill ranges through five districts, namely, Chittoor, Kadapa, Kurnool, Nellore, and Prakasam districts of Andhra Pradesh (UNEP WCMC 2017).

The species naturally occurs in the forest formation which is called as "5A/C3 Southern tropical dry deciduous forests" as per Champion and Seth (1968) classification and falls in the eco-terrestrial region IM1301 Deccan thorn scrub forests and as well as IM0201 Central Deccan Plateau dry deciduous forests (NBA 2016).

P. santalinus is limited to the forest areas which have been classified by Champion and Seth (1968) under Southern tropical dry deciduous forests (5A). But they are evidently predominant in a subtype forests, namely, "red sanders forests" (5A/C2). These species are found either in pure patches or in association with other species. In these forests red sanders is associated with other top story species, such as *Anogeissus latifolia*, *Kingiodendron pinnatum*, *Terminalia alata*, and *T. chebula*. *Chloroxylon swietenia* and *Erythroxylum monogynum* are typically present in the middle story.

Red sanders is also found in another subtype of eco-terrestrial region known as Southern dry mixed deciduous forests (5A/C3). Here, the tree species are largely associated with species like *Anogeissus latifolia*, *Cleistanthus collinus*, *Tectona grandis*, *Pterocarpus marsupium*, *Terminalia alata*, and *Kingiodendron pinnatum*. They also occur in an edaphic subtype of dry deciduous forests recognized by Champion and Seth (1968) in the component of *Hardwickia* forest (5A/E4). According to Champion and Seth (1968), the rock changes from slate and quartzite to gneiss support the growth of red sanders with little climatic changes, though the species disappears in adjoining areas red sanders forest (5A/C2). The trees are also recorded in the Saidapet division (Tamil Nadu), in the southern dry deciduous forests with its main associates as *Shorea tumbuggaia* and *Syzygium alternifolium* (Troup, 1983).

9.3 Population Status

P. santalinus is naturally distributed in Seshachalam, Veligonda, Lankamala, and Palakonda hill ranges around approximately 4.0 lakh hectares running through five districts, namely, Chittoor, Kadapa, Kurnool, Nellore, and Prakasam districts in Andhra Pradesh (NBA 2016). It grows on dry, hilly, often rocky ground and generally found at altitude ranging from 150 to 900 m. It prefers lateritic and gravelly soil and cannot tolerate waterlogging (Raju and Raju 2000).

Red sanders prefers mainly very shallow to shallow brown colored, sandy loam, or stony soils of friable nature (Raju and Srinivasulu 2008). Red sanders is not found in compact soils. Hot and dry climate, higher average temperatures, and low rainfall (500–800 mm) are general features of red sanders growing areas. These climatic factors coupled with geological formations and soil factors and may be some more unknown ecological factors form special "ecological niche" which is responsible for limited distribution of red sanders (IFGTB 2011).

Kodur plantation of 1865, raised on rich alluvial soil, showed, during the first 18 years, a mean crop height of 12.2 m, a mean girth b.h. of 45.4 cm, average number of stems per ha of 1512, and M.A.I. of 7.56 tonnes with bark or 7.38 tonnes without bark. The mean annual girth increment at the age was 2.5 cm. The M.A.I. is equivalent to 10.5 cum/ha. After the age of 18 years, the growth fell down, due to inadequate thinning (Troup 1983).

Raju and Nagaraju (1999) and Raju and Raju (2000) have studied the relation between geological formations and red sanders distribution and found that quartzite and shale formations are important for occurrence of red sanders. They reported that about 82% of red sanders area is covered mainly by Gulcheru, Nagari, Pullampeta, Bairenkonda, Cumbum, and Irlakonda quartzite and subordinately by Pullampeta and Cumbum shale (Raju and Nagaraju 1999; Raju and Raju 2000). It is also theorized that certain quartzites and shales formed during a short period in the Proterozoic era contained certain trace elements in significant quantities to favor the growth of red sanders.

As per the National Biodiversity Authority report 2016, red sanders plantations are found in four states, viz., Kerala, Karnataka, Tamil Nadu, and Andhra Pradesh, and natural populations are seen in Rajampet (S.R. Palem) and Tirupati in Andhra Pradesh.

Plantations of Kerala were planted in 1983 and they are 35 years old. The plantations of Tamil Nadu were located at Banavaram (Vellore division) and Malanthur (Tiruvallur division). The Malanthur plantation is 42 years old and Banavaram plantation is 34 years old. Plantations found in Andhra Pradesh were at Kodur (1956) (60 years old) and Narasingapuram near Tirupati which was about 30 years old. In Karnataka, a 29-year-old plantation is located at Jarakabande in Bangalore and 50 years old located at Shettihalli in Mandya division (NBA 2016).

The average growth of *P. santalinus* in different plantations/natural populations are presented in Table 9.1 (NBA 2016). The average tree height ranged from

Table 9.1 Growth of *P. santalinus* in plantation/natural populations

Location of plantation	Age in years	Total tree height (m)	Girth at breast height (cm)	Clear bole height (m)
Plantations				
Malanthur (TN)	45	15.13 ± 1.60	80.30 ± 13.25	3.73 ± 1.96
Banavaram (TN)	33	14.27 ± 1.83	102.67 ± 25.04	4.50 ± 2.78
Kodur (AP)	60	23.67 ± 4.27	100.33 ± 32.00	12.23 ± 3.49
Palappilly (Kerala)	33	17.23 ± 5.67	83.44 ± 25.49	6.18 ± 4.45
Kodanad (Kerala)	33	19.26 ± 3.14	96.43 ± 17.27	5.23 ± 4.19
Tirupati (AP)	28	15.54 ± 2.04	83.54 ± 25.49	5.59 ± 1.95
Jarakabande (Karnataka)	29	11.00 ± 0.92	67.11 ± 7.48	3.33 ± 1.08
Shettihalli (Karnataka)	50	10.95 ± 1.20	81.55 ± 11.50	3.43 ± 1.02
Natural populations				
Tirumala (AP)	–	9.38 ± 2.18	71.38 ± 9.92	3.56 ± 1.09
S.R. Palem (AP)	–	8.83 ± 1.83	76.18 ± 16.14	2.88 ± 1.39

Source: NBA (2016)

10.95 m (Shettihalli) to 23.67 m in plantations located at Kodur in Andhra Pradesh. The natural population at Tirumala in Andhra Pradesh recorded average tree height of 9.38 m. Among plantations 50-year-old Shettihalli plantation at Mandya recorded average height of 10.95 m which is less compared to other younger plantations. The 33-year-old plantations located in high rainfall areas at Palappilly and Kodanad recorded 17.23 and 19.36 m, respectively.

The average girth at breast height (GBH) varied from 67.11 cm (in 29-year-old plantation at Jarakabande, Bangalore) to 102.67 cm in 33-year-old plantation at Banavaram (TN). The 45-year-old plantation at Malanthur (80.3 cm) and 50-year-old plantation at Shettihalli (81.55 cm) recorded relatively lower GBH compared to some of the younger plantations, viz., 33-year-old plantation at Palappilly (83.44 cm), Kodanad (96.43 cm), and 29-year-old plantation at Narasingapuram (83.54 cm). The 60-year-old Kodur plantation recorded average GBH of 100.33 cm (NBA 2016).

Clear bole height (CBH) is the branch-free stem height which is correlated with yields and qualities of merchantable logs or timber result from branch shedding, a process influenced by numerous environmental factors especially crowding or stocking (Saha et al. 2012, 2014; Kuehne et al. 2013; Hardiansyah et al. 2015) and by genes (Lowell et al. 2014). The average clear bole height varied from 2.88 m in S.R. Palem natural population to 12.23 m in Kodur plantations. The Kodur plantation had unusually high clear bole height, as some of the trees had as high as 18.0 m clear bole height. Generally, the trees occurring in natural populations had lower clear bole height. However, some of the plantations at Shettihalli, Jarakabande, and Malanthur had less than 4.0 m average clear bole height.

Table 9.2 Population status of red sanders

Sl. No.	Division	Range	Total area of the range (ha)	Red sanders bearing forest area (ha)	Ref.
1.	Tirupati Wildlife	Chamala	22,938	11,100	Hegde et al. (2012)
2.		Tirupati	25,486	23,697	Hegde et al. (2012)
3.		Balapalli	28,673	28,628	Hegde et al. (2012)
4.	WLM Tirupati	–	68,738	58,000	Bhagyaraj (2017)
5.	WL Chittoor East	–	22,500	19,500	Bhagyaraj (2017)
6.	Rajampet	Chitvel	40116.46	25,781	Hegde et al. (2012)
7.	Rajampet	Rajampet	28202.37	1930	Senthilkumar et al. (2015)
8.	Kadapa	Siddhout	34691.83	26,808	Hegde et al. (2012)
9.		Vontimitta	16170.50	12,885	Hegde et al. (2012)
10.	Proddatur	Proddatur	10630.47	8740	Hegde et al. (2012)
11.		Badvel	39719.48	24,680	Hegde et al. (2012)
12.	Nellore	Rapur	19176.00	3000	Hegde et al. (2012)
13.		Atmakur	16766.71	2800	Hegde et al. (2012)

Population status of red sanders recorded by different authors in different places and time are mentioned in Table 9.2. According to Bhagyaraj (2017), Wildlife Management (WLM) Tirupati is bearing 58,000 ha of red sanders. Rajampet range is small red sanders bearing area of 1930 ha.

A study on the population structure of red sanders forest of Sri Lankamalleswara Wildlife Sanctuary (SLWS) was conducted by Ankalaiah et al. (2017); the range of occurrence is 165–246 individuals per ha with size class structure of individuals more confined to lower GBH classes with 364 individuals (39.5%) in 30–50 cm GBH class and 420 individuals (45.6%) in 51–70 cm GBH class totaling to 784 individuals (85.1%). The higher 71–90 cm GBH class has 129 individuals (14%), and only 7 individuals are present in >90 cm GBH class.

In view of the declining population of red sanders, finally in the year 2008, the Government of India amending its act gave permissions to cultivate red sanders which has opened doors for individuals and cooperate bodies to relish the rich outcome of cultivating this wild crop.

9.4 Growth Habit

P. santalinus is a deciduous tree growing as a small- to a moderate-sized with an erect clean bole having a rather dense, rounded crown. Under natural conditions *P. santalinus* attains an average girth of 1.5–1.9 m and a height of 9–11 m with a clean bole of 4.5–6.0 m. But under plantation conditions on rich soil, it has the capacity to reach a height of 15–18 m with a clean bole of about 9 m. As per the forest records of Kadapa district in Andhra Pradesh, large-sized trees up to 2 m girth are known to have existed in the past, which is proved by the stumps seen.

Studies on *P. santalinus* of 20 and 45 years old plantations of Karnataka revealed the growth variability of 130 trees in Jarakabande 'A' block varied from 25–79 cm. Further, the measurements of 98 trees in Hulikere plantation showed the variation from 37 to 111 cm with average girth of 73.83 cm. Soil parameters like pH, EC, and nutrients also showed favorable conditions for their growth (Kumar 2011).

9.5 Geoclimatic Factors

9.5.1 Climate

It is a strong light demander and does not tolerate overhead shade and is fairly fire-hardy. These trees are planted in hot and dry areas of South India, since the species are resistant to drought. In its natural habitat of Andhra Pradesh, especially in Kadapa, the temperature ranges from 37.8 to 46.1 °C. But during December, the mean daily temperature falls down to 37.8 °C, while the absolute lowest temperature also reaches to 11.7 °C. The region receives about 88.5% of the rains during June–November. In the Kadapa region, it rains for 46 days in a year with mean annual rainfall of 768 mm. But the relative humidity varies from 32% to 79% with an average being 58% (Troup 1983). It regenerates well in dry hot climate and requires rainfall ranging from 800 to 1000 mm annually for good growth. *Pterocarpus santalinus* is a light-demanding small tree, growing to 10 m (33 ft) tall with a trunk 50–150 cm diameter. It is fast-growing when young, reaching 5 m (16 ft) tall in 3 years, even on degraded soils. It is not frost tolerant, being killed by temperatures of −1 °C (Reddy 2018).

9.5.2 Topography

P. santalinus prefer an elevation of 150–900 m above sea level with dry, hilly often rocky terrains for their better growth. The species also prefer sloppy terrains to avoid waterlogging, and occasionally they are found on precipitous hill slopes. The trees are also found in the lower altitudes, but they show inclination to restrict themselves to cooler northern and eastern aspects (Troup 1983).

9.5.3 Geology and Soil

P. santalinus propagates well on rocky hills consisting of quartzite, shale, limestone, and laterite soils. But better tree growth is observed particularly well on lateritic loam. *P. santalinus* is not exacting in its soil requirements and is found even in poor shallow soil. Nevertheless, the best growth is noticed in rich soil, and the soil is generally shallow and poor due to fires and excessive grazing. The tree requires perfect drainage and is found mainly on stony or gravelly soil. The species cannot stand stiff in waterlogged soil, though it has been planted with success in rich alluvial ground, as in the well-known Kodur plantations of 1865 (Troup 1983). Well-drained red soils with graveled loam are suitable for the cultivation of red sanders trees. Red sandalwood grown on the shale subsoils, at altitudes around 750 m (2460 ft), and in semiarid climatic conditions gives a distinctive wavy grain margin (Reddy 2018). Trees need to be raised in a weed-free environment for first 2 years to allow them to fully develop without having to compete for nutrients (Reddy 2018).

The seeds retain their viability for about 1 year under normal storage conditions and are the best propagation material. Fruits are collected in March for the extraction of seeds. Seedlings or stumps prepared from 2-year-old nursery seedlings are used for planting in the field.

9.6 Morphology

Bark is blackish brown in color resembling the skin of the crocodile with rectangular plates of deep vertical and horizontal cracks. The bark is generally 1.3–1.8 cm thick with pale yellow or white blaze with numerous pink streaks exuding a copious red gum.

Locally, two different types of the tree are documented, with comparatively smooth bark and lighter colored heartwood while the other with rougher bark and darker colored heartwood. They are entitled as "female" and "male" trees, respectively, by the local woodcutters. But no botanical difference has been found so far, and these trees are found beside each other (Troup 1983). The wood of most trees has a normal grain, and there is also "wavy grain" which is rare.

9.7 Phenology

The study of variations in response to climate, seasons, and weather in plants for nature's rhythm from time to time is referred as phenology. The phonological observations help in the better understanding of interactions between meteorological variables and associated biological responses. This includes development of tree characteristics like flowering, leaf unfolding (or budburst), seed set and dispersal, and leaf fall in relation to climatic conditions (Davi et al. 2011) at different distinct periods and seasons. In the dry tropics, vegetative phenology varies widely with tree characteristics and soil conditions. The influence of global climate change has altered the climate of the regions with alterations in tree's characters.

The maturity period for red sandalwood trees is 12–15 years. *P. santalinus* shed their leaves during middle of January to middle of March and completes shedding in the range of 20–30 days and remains leafless for 35–40 days. In moister localities the leaf fall commences later in February and is complete by about the middle of March. The tree is the first to throw out a new flush of leaves, even though the species sheds its leaves earlier than most of its associates. End of March to early April, before any other of its associates has come into new leaf, red sanders is seen putting out tender, green leaves. The species remain leafless for a short period of about half a month in drier localities while almost a month in moister localities (Troup 1983).

The yellow flowers in short racemes appear with the arrival of new leaves during April and extends till middle of May to June. The racemes of the flowers are with 25 + 4 members, pedicellate of 16 mm long, bisexual, zygomorphic, and mildly odoriferous. Pollinator activity is limited to moonlit nights and early morning hours. According to Dayanand and Lohidas (1988), red sanders matured pods are formed after 11 months from their flowering. The natural fruit set is very low for about 6% in proportion to the quantum of flowers. The xenogamous fruits alone are carried to maturity and are dropped to autogamous and geitonogamous. Pods form rapidly but ripen only during next February–March when the tree has again shed its leaves, showing bunches of dark brown pods on most of its bare branches. The pods are retained in the tree till May, where the southwest winds help in blowing off and dispersing. The trees' pods are not only found clinging to the tree till hot season but are found beside new green pods for the present season (Troup 1983).

9.8 Natural Regeneration

Natural regeneration is through seeds. The seeds rejuvenate under their favorable conditions like loose soil, enough moisture, and free from forest fire and animal grazing. In natural reproduction, conditions supporting for growth are presence of loose soil where pods become wholly or partly embedded, shelter from the warmness of the sun during germinating and early seedling stages, and protection from fire and animals at advanced stages. Generally, sprouting starts during the early showers with the arrival of southwest monsoon. Under natural conditions, difficulty arises with the survival of seedling or sapling from abiotic and biotic factors. The restricted grazing and fire protection with growth in grass and other associates help in the development from seedling to sapling stage. At the later stages of development, species resist fire along with it's associates. But the seeds fall during the rest half of the fire season, so that the loss of sprouting seeds is minimized. Pod's fibrous tough coat prevents the seed from ground fire and light. Natural regeneration occurs effortlessly in protected areas from overgrazing and fire as the seeds begin to germinate after the initial monsoon showers (Troup 1983).

The seedlings of the natural forest for species like *P. marsupium* and *P. santalinus* exhibit dying conditions year after year during seedling stage, when the tree is struggling to develop a thick carroty root which can distribute a large vigorous and strong shoot, capable of resisting the drought, fire, and artificial grazing (Troup 1983).

In the region of top Pitchanur in Chittoor division, a study on the effect of a judicious and gradual opening of the overhead canopy with a view to establish natural seedlings was carried out during 1936. It was found that trees were of the height 23–46 cm and are found in patches with vigorous seedlings and broad healthy foliage and responded remarkably to gradual opening of the canopy. During the initial stages, the stumpy cover is removed, with which canopy is opened and lifted till the density is reduced to 0.4 (Troup 1983).

According to the study by Khurana and Singh (2001), ecology of tree seed and seedlings on tropical forest has great influence on abiotic factors like distribution, amount of rainfall, light, temperature, soil nutrient regimes, intensity of predation, and disturbance. Germination and seedling growth are vital for community and developing strategies for the conservation and restoration. But, rare species in the community are at high risk at local extinction due to these factors. The study on tree diversity in a sal has fair regeneration with gradual decrease from seedling, sapling, regeneration, and adult stages due to abiotic and anthropogenic interference by the dwellers (Shankar 2001). Similar results were attributed to red sanders regeneration as the number decreases from seedlings (276–369), saplings (78–99), and regenerating trees stages (49–77). The bottleneck progression from seedling to adult tree stage has greater impact on tree biomass and population (Ankalaiah et al. 2017).

The first mark of germination is seen after a week from sowing, where the development of the radicle pushes the pod upward for evacuation. The radicle pushes its way out of the hypocotyledonary portion in the form of a loop, through the stylar insertion in the pod, thereby creating a slit along the suture near the insertion. The loop then unbends, and the radicle enters the soil and develops into a taproot, while at the same time, the hypocotyledonary portion tends to become erect and, in this attempt, frees the cotyledons which by this time have half come out of the pericarp, through the slit already made. Sometimes, particularly when two seedlings emerge from one pod, the pod is raised above ground, falling with the expansion of the cotyledons; in other cases, the pod is left on or in the ground. The testa of the seed is always left within the pod. The cotyledons then spread out. The plumule begins its development and puts out the first tiny leaf with two very small stipules at its sides. The taproot which is now distinct readily grows in length and develops minute rootlets. It should be noted that though eventually the leaves are trifoliately compound, the first four or more leaves during the first season growth are simple leaves (Troup 1983).

Extrinsic factors like weather conditions, seed predation, and the capability to offer resources for juvenile fruits and seeds for their maturity by maternal parent play a vital role in the growth of trees (Stephenson and Bertin 1983). With reference to Chandrashekara (1996), among the tested pods of red sanders resulted in 0–75% were empty pods and 80–100% were dead and diseased pods with his observations on seed size variations. The observations on different pre-sowing treatments for germination of red sanders pods in the samples tested showed wide variation from 0% to 33.33%. This was probably due to the presence of 0–75% empty pods and 8–100% dead and diseased pods. The seeds of the red sanders recorded a very low percent of germination collected from different parts of Andhra Pradesh. The percent germination ranged from 6% to 10% in Anantapur and Kadapa and Tirumalai hills of Tirupati, respectively. The seeds of *P. marsupium* showed 68% germination rate when pods were soaked in 10% sulfuric acid for 10 min. The study was also on the variability in viable seeds from different locations (Sharma et al. 2000).

The species primary long root is with moderately thick terete, tapering, wiry to woody, lateral roots moderate in number, fibrous, and distributed down the main root with nodules. The hypocotyl is distinct from roots with its length of 1.0–2.5 cm, slightly compressed, and fusiform or tapering up or down with white turning green. The cotyledons subsessile or with petioles are up to the length of 1.5 mm along with foliaceous and fleshy membrane of 2–2.5 cm by 8–12 mm. They are unequally ovate, oblong entire, apex rounded, and with base rounded or sagittate or semi-sagittate. The natural seedlings have erect stem, wiry green pubescent, and with internodes of length 0.8–2.5 cm. The tree forms subsequent trifoliate leaves, with alternate and petiolate leaves during the first season. However, during later stages, linear-lanceolate or subfalcate stipules are formed which are pubescent and have acuminate tips. Leaves 3-foliolate, rarely 4-5-foliolate, leaflets ovate-orbicular or oblong, 4–6 × 3.5–5 cm, base obtuse-subcordate, margin entire, apex emarginated; petiole to 4 cm long, channeled above, petiolule to 6 mm. (Troup 1983).

The tree is said to regenerate well from coppicing, and a 40-year coppice rotation is said to be practiced in India (Kumar 2011). Propagation techniques in *P. santalinus* are through seeds, cuttings, and grafting. The details of each propagation are given below.

9.9 Seed Pretreatment

About 1 kg seeds are needed to raise seedlings for 1 ha of plantation. Seeds show only about 50–60% germination and about 40% plant survival. Seed treatment with GA_3 (gibberellic acid) + BA (benzyladenine) at 250 PPM (parts per million) helps in early and optimum germination. Alternatively, seeds can be soaked in cold water or in cow dung slurry for 72 h (Reddy 2018).

9.9.1 Raising Propagules

March, April, or May months are suitable for raising nursery from seeds. Manure and healthy pods are chosen to collect the seeds; these are sown in raised nursery beds. It takes 10–15 days for the germination to complete. Pretreated seeds (10–12 kg) are sown per bed (10 × 1 m) of sandy loam or loamy soil in April and covered with a thin layer of soil or hay. After 1 year, stumps can be prepared, which are planted in polybags in February to March, to be utilized after 4–5 months as stump-sprouted seedlings (http://vikaspedia.in/agriculture/crop-production/package-of-practices/medicinal-and-aromatic-plants/pterocarpus-santalinus?commenting).

9.9.2 Propagule Rate and Pretreatment

About 1 kg seeds are needed to raise seedlings for 1 ha of plantation.

Patel et al. (2018) treated red sanders pods for 11 different pre-sowing treatments to check their germination and survival percentage. The treatment with GA_3 with 500 ppm for 1 day performed well with survival percentage of 63.33% (7.67 days) and followed by the treatments with GA_3 with 250 ppm for 1 day, and seeds separated from the pods were directly sown in the poly bag without any treatment and showed survival percentage of 56.67% (9 days) and 60% (9 days), respectively. But the no treatment pods showed low survival percentage (40%) in comparison with other treatments.

Pods of the tree attain harvestable maturity in 10 months after pod set in the month of December and they take 45 days for growth and development during March-April. Though the pods mature in the month of December, it is better to harvest during starting of March as the moisture content is less. The average count of pods per kilogram was 1450 for small (0.65 g/pod), 1050 for medium (1.05 g/pod), and 650 for bigger (1.45 g/pod) pods. Studies on pre-sowing treatment of pods revealed that pods soaked in water for 3 days were found better for early germination (12.1 days) as compared to delayed germination (16.0 days) in case of concentrated H_2SO_4 treatment for 10 min. The months of March and February were found better as they germinate early (12.1 and 12.2 days) with maximum normal seedlings as compared to other months (Kodigowda 2002).

Among the different growth regulator treatments tried to enhance the germination rate, GA_3+BA at 250 ppm was found better as it produced maximum germination (48.75%) in minimum number of days (11.87 days) with maximum normal seedlings as compared to the treatments of GA_3, BA, GA_3+ BA, and thiourea at various concentrations. Added to this, treating with GA_3+BA at 250 ppm produces maximum number of leaves and leaf area along with shoot and root lengths and maximum fresh and dry weights of shoot and root of seedling. Higher percentage of one seeded pods (45.66%), 100 pod weight (113.66 g), 1000 seed weight (115.00 g), and germination percentage (34.33%) with minimum of empty pods (21.00%) were recorded in pods of isolated tree as compared to plantation tree (24.00%, 108.66 g, 99.83 g, 22.00%, and 41.00%, respectively). Two seeded pods were comparatively more (3.00%) in plantation tree than in isolated tree (Kodigowda 2002).

Generally, the dry pods are collected during the months February–June. Then they are dried in sun for 3 days and stored in bamboo baskets or gunny bags till required. Dry pods weigh for about 900–1400 per kg. Germination occurs in about a fortnight. According to the studies, germinative capacity is between the range 10 and 12% in Andhra Pradesh, but in Orissa it is reported as 80% and in Chingleput, Tamil Nadu, is reported with the germinative range of 25–30% (Troup 1983).

Pods are collected with their wings directly from the trees for regeneration. The germinative capacity decreases as the seeds are removed from the pod, and according to the study by Dent (1948), there is no loss of viability for at least 8 months. But, the seed viability is lost if stored for more than 1 year. In few regions of Tamil Nadu, especially in Chengalpattu division, the pods are immersed in cow dung slurry before sowing for 48 h, and pods are also immersed in cold water for 3 days to increase the germinative capacity (Troup 1983).

9.9.2.1 Nursery Technique

Red sanders nurseries at Andhra Pradesh are prepared with well-drained soil and small pebbles or gravel in loamy or sandy loam soils mixed with farmyard manure. The soil is broken into powder from lumps, and they are kept dry up to 30 cm deep. The aboveground bed is prepared of 12 × 1.2 m with periphery supported by slit bamboos or other locally available materials. The pretreated seeds of approx 10–15 kgs are sown per bed with a thin layer of earth or hay. Generally, the top layer is covered with dry leaves or straw and watered profusely. About 7–14% of the plants are obtained with 1000–2000 seedlings per bed (Troup 1983).

In Tamil Nadu, especially in the region of Chengalpattu, the seedlings of height 4–10 cm are transplanted into polythene containers, and the containers are kept under shade and regularly watered for a month. Later the plants are further watered for 3–4 months with the removal of shade. During the end of the period, carroty roots are developed with its height of 20–25 cm. In Orissa, pods are sown into pot containers mixed with soil and farmyard manure and regularly watered till monsoon season. By this method, 60% plants are reported to be obtained (Troup 1983).

This plant can be raised by direct sowing, in nursery beds or by stumps. Stump planting gives highest percentage of results. Two-year-old seedling gives best results when transferred to the field. It is a highly shade-intolerant plant species; it requires plenty of sunlight for proper growth. Large-scale plantations of this tree species have been established in recent years in Andhra Pradesh.

The seeds retain their viability for about 1 year under normal storage conditions and are the best propagation material. Fruits are collected in March for the extraction of the seeds. Seedlings or stumps prepared from 2-year-old nursery seedlings are used for planting in the field.

9.10 Nursery Rising

March, April, or May months are suitable for raising nursery from seeds. Mature and healthy pods are chosen to collect the seeds; these are sown in raised nursery beds. It takes 10–15 days for the germination to complete. Pretreated seeds (10–12 kg) are sown per bed (10 × 1 m) of sandy loam or loamy soil in April and covered with a

Fig. 9.1 Red sanders nursery. (Courtesy Reddy 2018)

thin layer of soil or hay. After 1 year, stumps can be prepared, which are planted in poly bags in February–March, to be utilized after 4–5 months (Reddy 2018) (Figs. 9.1, 9.2, and 9.3).

9.11 Land Preparation and Fertilizer Application

The land is plowed and harrowed repeatedly, and the soil is brought to a fine tilth. Pits of size 45 × 45 × 45 cm are dug at a spacing of 4 × 4 m. The pits are filled with topsoil mixed thoroughly with 10–15 kg FYM (farmyard manure) and 10 g linden dust to protect the planting stock from soilborne fungi (Reddy 2018).

9.12 Transplanting and Spacing

The best time for planting is end of May–June that is the onset of rainy season. Generally, the stump-raised seedlings or stumps obtained from 2-year-old nursery plants are transplanted. A plant population of about 600 per hectare is recommended with a spacing of 4 × 4 m. These also can be planted near the boundaries of farm houses, farmlands, and gardens (Reddy 2018).

Fig. 9.2 Stumps in poly bags (Reddy 2018)

Fig. 9.3 Five-year-old red sanders plantation. (Courtesy Reddy 2018)

9.13 Intercropping System

No intercropping system has so far been studied in this crop; however, herbaceous rhizomatous species may be grown as an intercrop in inter-row spaces (Reddy 2018).

9.14 Maintenance Practices

About 10–15 kg FYM per plant per year and 150:100:100 g NPK (nitrogen, phosphorus, and potassium) per plant per year are required for at least 5 years. Fertilizer should be applied in 15- to 20-cm-deep circular trenches dug around the plant at a distance of 60 cm. Full dose of P and K along with one-third N should be applied at the end of February. The remaining N should be applied in two split doses during June–July and October–November (Reddy 2018).

Application of inorganic fertilizers should always be followed by irrigation in two split doses during June–July and October–November. Gap filling is done 1 month after planting. Weeding is done manually as and when necessary and particularly just before manuring. The soil around the basin is loosened frequently by hoeing (Reddy 2018).

9.15 Varieties

In nature, two types of red sanders trees are observed—wavy-grained and straight-grained. The wavy-grained wood is more in demand in trade and is preferred for commercial plantation. No commercially released varieties are available.

9.16 Vegetative Propagation

Red sanders is conventionally propagated through seeds and vegetative methods. Propagation through seed is often very difficult because of a hard seed coat coupled with poor viability (Dayanand and Lohidas 1988; Naidu and Rajendrudu 2001). In India, several studies on propagation of *Pterocarpus santalinus* through seeds and other modes of vegetative propagation have been attempted, but not much success rates are found till today; thus there is a need for developing a propagation method for red sanders to meet the demand.

Rooting of cuttings in different seasons with growth regulator treatment was a failure with all the types of cuttings and in different seasons, except for callus formation. However, the hard and semi-hardwood cuttings responded well to sprouting with the treatment of growth regulators at lower concentrations of IBA and NAA during the months of October and February. The sprouted cuttings have dried after 30 days of treatment (Kodigowda 2002).

Variations were observed in biochemical components of hard-, semi-hard-, and softwood cuttings. Generally, the carbohydrate content was higher in hardwood cuttings, and its content was highest (16.00%) in February compared to other months of treatment. The lowest carbohydrate content (14.15%) was recorded in softwood cuttings during October after the growth regulator treatment (Kodigowda 2002).

The development of vegetative propagation techniques represents the first step in the process of domestication of a tree species (Leaky and Simons 1998). Vegetative propagation offers a unique opportunity of avoiding the problem of recalcitrant seeds predominant in tropical tree species. It also facilitates the transfer of the genetic potential as well as the nonadditive variance of the parent to the new plant (Puri and Khara 1992).

Clonal propagation is possible through stem cuttings collected from matured trees and is one of the most effective tools for improvement in forestry (Kala and Kumaran 2015). For improvement in propagation techniques, many plant growth regulators with different concentrations have been investigated and optimized.

Lower concentrations of plant growth hormones, viz., IAA, IBA, and IPA, were used, and 25% rooting was seen in propagation of stem cuttings of red sanders (Reddy and Srivasuki 1990). Since growth regulators have a higher impact on root and shoot induction, using such plant growth enhancer will help in propagation of red sanders. *Pterocarpus soyauxii* stem cuttings treated with 8000 and 4000 μg ml^{-1} had 100% rooting (Egbe et al. 2012).

The success of graft take was very poor (18%), and it shows incompatibility of rootstock and scion. The grafting required 20.33 days for graft union to take place, and the average number of leaves per sprout was 4.66 (Kodigowda 2002).

Reddy and Srivasuki (1990) have attempted grafting in red sanders. They have obtained an overall graft take of only 26% by in situ cleft grafting performed on 4-, 7-, and 10-year-old stocks. They have also observed the retention of wavy grains in scion wood with a little higher bark thickness. Fernando et al. (2000) worked on propagation using air layering, cuttings, and tissue culture in semi-hardwood and hardwood treated with hormones (NAA and IAA) and found that there was very poor formation of any roots in hardwood material, whereas root primordial was seen after 6–8 weeks in semi-hardwood material of *P. santalinus*.

Habou et al. (2017) worked on air layering of *P. erinaceus* indicating that regular supply of water with a weekly interval is required for root formation and its viability further its being a low cost producing technique. Egbe et al. (2012) indicated that IBA concentrations of 16,000 and 8000 μg ml^{-1} were good for rooting in leafy stem cuttings and found 100% rooting in *Pterocarpus soyauxii*. Kedharnath et al. (1976) have reported about 38.70% graft success during March followed by 27.72% during February in red sanders by cleft method of grafting.

Reddy and Srivasuki (1990) attempted macro-propagation of *P. santalinus*, by stem cutting, air layering, and grafting indicating that there was incompatibility exhibited in grafting but observed the retention of wavy grains in scion wood with a little higher bark thickness and high degree of success in air layering, but it had got its own limitations; the use of stem cuttings treated with suitable growth regulators with suitable concentrations showed success in rooting by 25%.

9.17 Plantation Practices

9.17.1 Land Preparations and Fertilizer Application

The land is plowed and harrowed repeatedly and the soil is brought to a fine tilth. Pits of size 45 × 45 × 45 cm are dug at a spacing of 4 × 4 m. The pits are filled with topsoil mixed thoroughly with 10–15 kg farmyard manure and 10 g lindane dust to protect the planting stock from soilborne fungi.

9.17.2 Transplanting and Optimum Spacing

The best time for planting the crop in the field is end of May–June which is the onset of rainy season. The stump-raised seedlings or stumps obtained from 2-year nursery plants are transplanted. A plant population of about 600 per hectare is recommended with spacing.

9.17.3 Intercropping System

The herbaceous rhizomatous species can be grown as an intercrop in inter-row species.

9.17.4 Interculture and Maintenance Practices

About 10–15 kg of farmyard manure per plant per year and 150:100:100 g NPK per plant per year are required for at least 5 years. Fertilizer should be applied in 15–20-cm-deep circular trenches dug around the plant for a distance of 60 cm. Full dose of P and K along with one-third N should be applied at the end of February. The remaining N should be applied in the two split doses during June–July and October–November. Application of inorganic fertilizers should always be followed by irrigation. Gap filling is done 1 month after planting. Weeding is done manually as and when necessary and particularly just before manuring. The soil around the basin is loosened frequently at hoeing.

9.17.5 Irrigation Practices

The plants are irrigated immediately after transplantation. And further irrigation is done on alternate days up to 15 days. After the seedlings get established, irrigation can be done at an interval of 10–15 days, depending on weather conditions (http://vikaspedia.in/agriculture/crop-production/package-of-practices/medicinal-and-aromatic-plants/pterocarpus-santalinus?commenting).

The species of *P. santalinus* are grown well in the methods of raising forest plants by direct sowing, entire planting, and stump planting. The stump planting showed highest percentage of survival (87%), as against 20% for direct sowing and 13% for entire transplanting, with respect to the study in Tamil Nadu during 1939–1940. Two-year-old nursery-raised plants of stumps are comparatively better than 1-year-old nursery plants. The plants also grow moderately well through cuttings (Troup 1983).

In Andhra Pradesh, the seeds are directly sowed in mounds, whereas in Tamil Nadu, the patches of cultivation are burnt, and trees are spaced between 2 m^2 and have four weeding in a year for better results. The present practice is with the pit size and spacing of 3 × 3 m in 30 m^3. The pits are made in the cultivating area by the end of May, and planting is carried with the onset of monsoons. One or two hoeings round the plant with 1 m radius and 15 cm depth are placed for the first 2 years during the cultivation (Troup 1983).

The plantation raised by H.H. Yarde, from 1865 to 1869 in Kodur, Kadapa district of Andhra Pradesh, was planted by basket transplantation. The seedlings are raised by seed sowing in nursery from July where the saplings are transplanted without disturbing the soil round the roots, into bamboo baskets of 30 cm deep and 24 cm diameter. These baskets are watered every 2 or 3 days and are kept under shade. The baskets are buried in pits of about 2.4 m apart and are watered till monsoon rains, when plants recovered from the shock due to transplantation. For the initial few years, plants are regularly weeded with sheltering against hot sun, by fixing leafy branches around each plant at top (Troup 1983).

In the study carried out in southern Kadapa division, planting nursery-raised seedlings and direct-sowed seeds responded with equal good results. The direct sowing is spaced with 1.2 m^2, and standard method for raising plantations was followed in that division. The existing method in Andhra Pradesh was by stumps of 1–2-year-old nursery stock as planting material, but larger and older stock of stumps is preferred (Troup 1983).

9.18 Injuries and Protection

9.18.1 Insects

Red sanders are vulnerable toward defoliators, more likely during summer especially prone to Pyralidae (Lepidoptera). The damage is reduced with the first heavy monsoon rains. Caterpillars attack red sanders poles mostly between 30 and 38 cm breast girth. The form of damage may be serious as it apparently eats down to the cambium layer. The attack generally begins at the fork of a tree. The larval period of activity closes by about April in Chittoor.

9.18.2 Disease and Pest Control

Leaf-eating caterpillars have been found to damage the crop during April–May. These can be controlled by spraying 0.2% monocrotophos twice at weekly intervals.

9.18.2.1 Plants

Young plantations are often badly affected by climbers, especially *Cassytha filiformis* which completely entwines the young seedlings and kills them in a matter of months. These have to be controlled by physical removal or by allowing sheep browsing selectively.

9.18.2.2 Animals

Sambar and spotted deer, particularly the former, greedily browse seedlings of red sanders in natural forest and plantations. In rab patches where red sanders is grown in the center, it is an advantage to raise *Cassia siamea* on the border as a protective barrier, as red sanders seedlings, even when partially browsed, have benefitted from such protective barrier.

Domestic cattle are responsible for browsing, apart from the damage by soil compaction. But grazing also keeps down the heavy grass and reduces the risk of fires. According to a study, on an average 20–40% of the seedlings/regenerating plants are lost due to cattle grazing and other natural phenomena.

9.18.2.3 Anthropogenic Pressure

In the past red sanders forests have suffered heavily due to overexploitation. This is a commercially valuable species and has suffered both from lawful and unlawful commercial overexploitation. Not only the stem but also the roots were extracted in the past, and the tree has escaped total extermination due to its wonderful adaptation and coppicing power.

9.18.2.4 Fires

Normally the fires creep along the ground burning grass and fallen leaves, but sometimes they cause permanent injury to saplings and poles while completely destroying regeneration to ground level. It has not been found possible to diminish the fire occurrences. Every year large areas are burnt and the forest floor laid bare. The study by Sridhar illustrates the effect on forest fire in the southern Eastern Ghats bearing red sanders tree where 60% of the saplings were found to be lost due to fire.

9.19 Yield

Pod yield from 15-year-old trees is 30 tons per hectare. Heartwood yield per tree after felling is 250 kg. Thus, 15,000 tons per hectare of heartwood is expected after 15–20 years. The estimated cost of raising and maintaining the crop is about Rs.55000 for 5 years. At an average market rate of Rs.75/kg, an income of Rs

112.5 lakhs is expected since it is a wild plant and the tree takes at least 30–40 years to come to the stage of cutting when it is grown under wild conditions.

9.20 Silviculture Characters

Some of the important silvicultural characteristics of *P. santalinus* are as follows:

- *P. santalinus* is a light demander and will not tolerate overhead shade.
- It requires abundance of growing space for proper crown development.
- Regular thinning on a 5-year cycle is required to promote the best growth in plantations.
- Resistant to fire than many of its associates.
- It shows the phenomenon of annual dying back.
- It is an excellent coppice when stumps are up to 1.2 m girth.
- It produces root suckers freely.
- They grow well in lateritic and gravelly soils and avoid waterlogging.
- The plantations require regular weeding during the first few years.
- Dying and misshapen seedlings and saplings are cut back to the ground with a sharp instrument.
- Seedlings are freed on rank weed growth by clearing a 60 cm diameter space round each.
- Climber cutting is carried thoroughly, and all red sanders trees are freed of these.

References

Ahmed M, Nayar MP (1984) Red sanders tree (*Pterocarpus santalinus* Linn.f.) on the verge of depletion. Bull Bot Surv India 26:142–143

Ankalaiah C, Mastan T, Reddy MS (2017) A study on the density, population structure and regeneration of red sanders *Pterocarpus santalinus* (Fabales: Fabaceae) in a protected natural habitat – Sri Lankamalleswara wildlife sanctuary, Andhra Pradesh, India. J Threatened Taxa 9 (9):10669–10674. https://doi.org/10.11609/jott.2869.9.9.10669-10674

Azamthulla M, Rajkapoor B, Kavimani S (2015) A review on *Pterocarpus santalinus* Linn.f. World J Pharm Res 4(2):282–292

Bhagyaraj A (2017) Red sanders invisible activities in Chittoor district: a case study. Int J Adv Res 5(9):1073–1080

Champion HG, Seth SK (1968) A revised survey of forest types of India. Govt. of India Press, New Delhi

Chandrashekara R (1996) Propagation studies of economically important tree species. MSc (Agri) thesis, University of Agricultural Sciences, Bangalore

Davi H, Gillmann M, Ibanez T, Cailleret M, Bontemps A, Fady B, Lefevre F (2011) Diversity of leaf unfolding dynamics among tree species: new insights from a study along an altitudinal gradient. Agric For Meteorol 151:1504–1513

Dayanand T, Lohidas T (1988) Effect of different treatments on pod germination of Red Sanders (*Pterocarpus santalinus* Linn. f). Indian J For 11:87–88

Dent TV (1948) Indian forest records. Silviculture. Vol. 7, No 1. Seed storage, with particular reference to the storage of seed of Indian forest plants. The Manager of Publications, Delhi

Egbe EA, Chuyong GB, Fonge AB, Tata BL, Tabot PT (2012) The effects of different concentrations of indole-3-butyric acid (IBA) on leafy stem cuttings of four tropical timber species. J Hort For 4(5):85–91

Fernando KCE, Oayatilake GA, Senaratne R (2000) Studies on the propagation of red sandal wood (*Pterocarpus santalinus* Linn.f.) an endangered medicinal tree species, Proceedings of the Sixth Annual Forestry and Environment Symposium of the Department of Forestry and Environmental Science. University of Sri Jayewardenepura, Sri Lanka

GOI (2014) Ministry of India. Ministry of Environment and Forest Publications. Red sanders. http://moef.nic.in/downloads/public-information/Chap-8-new.pdf

Habou R, Bationo BA, Laouali A, Segla KN, Adjonou K, Kokutse AD, Mahamane A, Kokou K (2017) Vegetative propagation by aerial layering of *Pterocarpus erinaceus*: in the Sudanian zone. Int J Recent Adv Multidiscip Res 4(10):2902–2908

Hardiansyah G, Ruslandi Hidayat D, Nugraha A, Mahardi M, Putz FE (2015) Clear bole lengths and crown characteristics of line-planted and naturally-regenerated *Shorea leprosula* trees: a case study from Kalimantan, Indonesia. J Trop For Sci 27(3):420–426

Hegde M, Singh BG, Krishnakumar N (2012) Non-detriment findings (NDFs) study for *Pterocarpus santalinus* L. f. (Red Sanders) in India. Available at https://cites.org/sites/default/files/ndf_material/NDF%20Report_Pterocarpus%20santalinus_India.pdf

IFGTB (2011) Collection of quantitative field data through rapid assessment of population, growing stock and natural regeneration status of *Pterocarpus santalinus* L.f. for CITES non-detriment findings study. A Final Technical Report of completed project submitted to MoEF Govt. of India during November, 2011, p 116

Jadhav SN, Reddy KN, Reddy CS (2001) Conservation assessment and management planning for medicinal plants of Andhra Pradesh. Medicinal plants conservation center (MPCC) and FRLHT. Hyderabad, pp 1–39

Kala S, Kumaran K (2015) Improved clonal propagation techniques for mass multiplication of *Bixa orellana* L. Indian For 141(3):279–284

Kedharnath S, Rawat MS, Uniyal DP, Lakshmikanthan D (1976) Studies on field grafting and the growth of the grafts in Red Sanders. Indian For 102(11):761–765

Khurana E, Singh JS (2001) Ecology of tree seed and seedlings: implications for tropical forest conservation and restoration. Curr Sci 80(6):748–757

Kodigowda K (2002) Morphometry and standardization of propagation techniques in red sanders (*Pterocarpus santalinus* L.f.). Division of Horticulture, University of Agricultural Sciences, Bangalore

Kuehne C, Kublin E, Pyttel P, Bauhus J (2013) Growth and form of *Quercus robur* and *Fraxinus excelsior* respond distinctly to initial growing space: results from 24-year-old Nelder experiments. J For Res 24:1–14

Kumar ANA (2011) Variability studies in *Pterocarpus santalinus* in different aged plantations of Karnataka. My For 47(4):343–353

Kumar ANA, Joshi G (2014) *Pterocarpus santalinus* (Red Sanders) an endemic, endangered tree of India: current status, improvement and the future. J Trop For Environ 4(02):1–10

Kumar S, Sane PV (2003) Legumes of South Asia, a check-list. Royal Botanic Gardens, Kew

Lakshminarasimhan P, Mondal MS (2012) A pictorial guide to some of the Indian plants included in the CITES and negative list of exports. Botanical Survey of India, Howrah, pp 22–24

Leaky RRB, Simons AJ (1998) The domestication and commercialization of indigenous trees in agroforestry for alleviation of poverty. Agrofor Syst 38:165–176

Lowell EC, Maguire DA, Briggs DG, Turnblom EC, Jayawickrama KJS, Bryce J (2014) Effects of silviculture and genetics on branch/knot attributes of coastal Pacific Northwest Douglas-fir and implications for wood quality—a synthesis. Forests 5(7):1717–1736

Mahammadh RV (2014) Smuggling of red sandal in India: a review study. Int J Econ Bus Rev 2 (10):55–63

Mulliken T, Crofton P (2008) Review of the status, harvest, trade and management of seven Asian CITES – listed medicinal and aromatic plant species. CITES

Naidu CV, Rajendrudu G (2001) Influence of kinetin and nitrogenous salts on seed germination of red sanders (*Pterocarpus santalinus* Linn. f.). Seed Sci Technol 2:69–672

National Biodiversity Authority (NBA) (2016) A study on comparison of natural and plantation grown red sanders wood. NBA, Chennai

Pandey B, Gangrale D, Upadhyay N, Tiwari P (2014) Physiochemical analysis of *Pterocarpus santalinus* L.f. extracts. Indian J Sci Res 4(1):201–204

Patel HS, Tandel MB, Prajapati VM, Amlani MH, Prajapati DH (2018) Effect of different pre-sowing treatments on germination of red sanders (*Pterocarpus santalinus* L.f.) in net house condition. Int J Chem Stud 6(2):876–879

Prasad RCP, Reddy CS, Raza SH, Dutt CBS (2008) Population structure, age gradations, and regeneration status of *Pterocarpus dalbergioides* Roxb., an endemic species of Andaman Islands, India. Pac J Sci Technol 9:658–664

Puri S, Khara (1992) Influence of maturity and physiological status of woody cuttings: limits and promises to ensure successful cloning. Indian For 118(8):560–572

Raju KK, Nagaraju A (1999) Geobotany of red sanders (*Pterocarpus santalinus*): a case study from the southeastern portion of Andhra Pradesh. Environ Geol 37:340–344

Raju KK, Raju AN (2000) Biogeochemical investigation in south eastern Andhra Pradesh: the distribution of rare earths, thorium and uranium in plants and soils. Environ Geol 39 (10):1102–1106

Raju KK, Srinivasulu S (2008) The red sanders forestry: an environmental appraisal towards an effective legislation. In: Proceedings of ecocity world summit 2008, San Francisco

Reddy OSK (2018) Red sanders – *Pterocarpus santalinus* – Rakta chandan – the red gold – green wealth – an opportunity for Indian farming community – let's explore. https://www.linkedin.com/pulse/red-sanders-gold-green-wealth-future-generations-indian-osk-reddy/

Reddy K, Srivasuki KP (1990) Vegetative propagation of red sanders (*Pterocarpus santalinus* L.). Indian For 116(7):536–540

Saha S, Kuehne C, Kohnle U, Bauhus J (2012) Growth and quality of young oaks (*Quercus robur* and *Q. petraea*) grown in cluster plantings in Central Europe: a weighted meta-analysis. For Ecol Manag 283:106–118

Saha S, Kuehne C, Bauhus J (2014) Intra- and interspecific competition differently influence growth and stem quality of young oaks (*Quercus robur* L. and *Quercus petraea* (Mattuschka) Liebl.). Ann For Sci 71:381–393

Senthilkumar N, Mayavel A, Subramani SP, Balaji K, Deenathayalan P (2015) Red sanders, *Pterocarpus santalinus* L.f. in Rajampet forest range, Rajampet forest division, Andhra Pradesh, India. Adv Appl Sci Res 6(10):130–134

Shankar U (2001) A case of high tree diversity in a sal (*Shorea robusta*)-dominated lowland forest of Eastern Himalaya: floristic composition, regeneration and conservation. Curr Sci 81 (7):776–786

Sharma A, Tiwari KP, Tiwari HC, Koshare M (2000) Pre-sowing seed treatments in *Pterocarpus marsupium* Roxb. to enhance seed germination and seedling growth. J Trop For 16(2):23–30

Stephenson AG, Bertin RI (1983) Male competition female choice and sexual selection in plants. In: Real (ed) Pollination biology. Academic, New York, pp 109–149

Troup (1983) The silviculture of Indian Trees. Volume IV. Leguminosae. Printed by The Manager, Government of India Press, Nasik, Published by The Controller of Publications, Delhi

UNEP WCMC (2017) Report on species/country combinations selected for review by the Plants Committee following CoP16 P 22–32

Vedavathy S (2004) Cultivation of endemic red sanders for international trade. Nat Prod Radiance 3 (2):83–84

Wenbin H, Xiufang S (2013) Tropical hardwood flows in China: case studies of rosewood and Okoume. For Trends Assoc, Washington, DC

Pests and Diseases of *Pterocarpus santalinus* 10

Umalatha and M. Anuradha

Abstract

The host of benefits forests provide are indispensable, and the flora and fauna of the forests are vulnerable to multiple threats. Apart from some abiotic factors such as fire, wind, storms, and climate change, there are many biotic factors like insects and diseases, which can have a devastating impact on the forest landscape. Man-made forests in general and particularly trees are susceptible to pests and diseases. Red sanders, a species with high timber value, is also affected; however the information on disease and pests is scanty. The tree is introduced in many geographical locations, and plantations are successfully maintained in various climatic zones. For sustainable utilization and to meet the global demand, the red sanders cultivation needs to be encouraged. For achieving healthy and resilient plantations, it is imperative to have a pest and disease management in place. In this chapter infections and infestations reported in red sanders are reviewed.

Keywords

Red sanders · Fungal infestation · Nematode infestation · Termite infestation · Heart wood borer

Umalatha (✉)
Padmashree Institute of Management and Sciences, Bangalore, Karnataka, India

M. Anuradha
Department of Biotechnology, Padmashree Institute of Management and Sciences, Bangalore, Karnataka, India

T. Pullaiah et al. (eds.), *Red Sanders: Silviculture and Conservation*,
https://doi.org/10.1007/978-981-13-7627-6_10

10.1 Introduction

Pterocarpus santalinus is in high demand in international market and is known for its gorgeous dark scarlet red wood with superlative technical qualities coupled with many medicinal properties. Due to overexploitation red sanders is endangered now, and this endemic tropical tree species from India is in need of restoration (Arunkumar and Joshi 2014). As this precious tree is endemic and threatened, great care should be taken for its survival and restoration. According to Luna (2005), forest fire causes extensive damage to young red sanders seedlings, sometimes wiping them completely. Young plantation of red sanders is also found to be badly affected by climbers and other competitive forest plants. This timber-yielding species has inherent resistance against attack by fungi, insects, and marine borers. Red sanders will not encounter any serious pest problem either in nurseries or in plantation. Due to changes in silvicultural and pre-harvesting practices, this property of high durability due to inherent resistance has become unreliable and reportedly interferes with their durability performance (Sundararaj et al. 2015). The main objective of this chapter is to provide an insight and better understanding of the pests and diseases encountered by *Pterocarpus santalinus.*

10.2 Fungal Infestation

Sankaran et al. (1984) reported leaf blight of *P. santalinus* caused by fungi *Sclerotium rolfsii*. Apart from *Sclerotium rolfsii*, there are many other fungi causing foliage blight in *P. santalinus*. *Colletotrichum gloeosporioides*, *Coniella fragariae*, *Phomopsis* sp., *Phoma glomerata*, and *Phoma eupyrena* are the important fungal pathogens which cause foliage infection in red sanders.

In Kerala during a routine forest survey in 1985, the pink disease caused by fungal pathogen *Corticium salmonicolor* was recorded in *Pterocarpus santalinus*. Rajasekhar et al. (1994) reported leaf spot disease in red sanders, and the causative agent is *Cylindrocladium scoparium*. This organism upon infection causes visible spots on the leaves of *P. santalinus*. During winter a severe leaf spot disease of red sanders was observed in Kadapa, Chittoor, and Nellore districts of Andhra Pradesh. The symptoms of the disease observed under field conditions were obvious pinhead-sized light-brown spots, water-soaked lesions which later turn dark brown, with central near whitish zone especially on the upper surface of the leaf. Premature leaf fall may occur during severe attack. Two or more spots may coalesce to form irregular patches which may involve major damage to leaves. At the center of the old lesions, shot holes may appear. However, infections were not noticed on the other plant parts (Rajasekhar et al. 1994). Reddy and Dayanand (1983) investigated the seed-borne mycoflora of red sanders (*Pterocarpus santalinus* Linn. f.). They reported *Aspergillus niger*, *A. flavus*, *Cladofiriwn cladosporioides*, and *Fusarium* sp. and affirmed seed infections due to these fungi in red sanders.

P. santalinus raised in root trainer nurseries of Kerala were found to be affected with foliage diseases. *P. marsupium* and *P. santalinus* were affected with foliage blight diseases under nursery conditions. In conducive edaphic and environmental factors, inoculum of most of the nursery pathogens activates in the presence of susceptible host. However, chances of seedling infection will be less even under conducive environmental conditions as inoculum potential of pathogens is considerably negligible in soilless medium employed in root trainer nurseries (Mohanan et al. 2005). Though root trainer nurseries used soilless or soil-free growing media maintained under hygienic conditions, infection by *Sclerotium rolfsii* in both the species is still recorded. Both the *Pterocarpus* species are highly susceptible to fungus, and *Sclerotium rolfsii* is a major cause for the spread of the foliage blight. Most of the soil-inhabiting, disease-causing fungi subsist mainly on the dead organic material in the presence of surplus, readily available nutrients in the organic compost, whereas soilless media are least attractive to the pathogens, and limited nutrients cause competition among them, and hence there should be less infections in soilless media root trainer nurseries. The compost prepared from forest weeds is the major constituent of the growing medium in root trainers, and it is suspected that sclerotium of the pathogen which is very resistant to environmental stress persisted in the compost and contributed to the disease (Mohanan et al. 2005).

10.3 Insect Pests

Fruit, seeds, leaves, and trunk of the fully grown trees in forest ecosystem are known to be attacked by numerous genera of insects. The pest infestation can cause devastating effects on and wipe of certain susceptible species. Though abundant population of *Eotetranychus sexmaculatus* (Riley) of family *Tetranychidae* was reported in Baruipur of West Bengal, there it had no damaging effect on *P. santalinus*. Sporadic infestation of *Planococcus* species of family Pseudococcidae was found in Gosaba of West Bengal. As it was least abundant, no noticeable damage was seen in the plant *P. santalinus* (Jash et al. 2018). It is to be noted that *P. santalinus* does not occur naturally in West Bengal, perhaps authors have the pests on trees of red sanders planted there. Evidently in mature stems and roots of *P. santalinus*, no serious insect pest and disease were observed. However, decrease in seed viability is observed and seeds are prone to seed borer. By proper drying (up to 12% moisture) and using carbon disulfide in storage, these seed borers can be controlled to some extent. Leaf-eating insects and white grub attack early growth stages of *P. santalinus* plants in nursery. These pests were controlled by spraying 0.003% endosulfan at fortnightly intervals and application of phorate 10 G near the root zone. Seed treatment with thiram at 3 g/kg of seed is essential to keep the plants disease-free in nursery and early stages of development in the field (Anonymous 2007). Caterpillars of Pyralidae (Lepidoptera) attack red sanders poles generally at the fork of the tree. Defoliators attack red sanders during summer, but damage is reduced during heavy monsoon. Leaf-eating caterpillars damage *P. santalinus* crop during April-May and can be controlled by spraying 0.2% monocrotophos twice at weekly intervals (National Medicinal Plants Board 2008).

10.4 Nematode Infestation

Nowadays in forests, diseases caused by nematodes are recognized as a significant problem, and it is a density-dependent one and becomes visible when nematode population exceeds critical threshold of economic damage. Limited information is available on diseases and nematode population dynamics in forests of Tamil Nadu where red sanders are naturally distributed. Study conducted for a period of 1 year at Sennamalaikaradu, Mettupalayam, and Coimbatore revealed the presence of soil and root population of parasitic nematodes, viz., *Helicotylenchus dihystera*, *Hoplolaimus seinhorsti*, *Meloidogyne incognita*, and *Tylenchorhynchus mashhoodi*, in red sanders plantations (Sivaprakash et al. 2009a, b). In this area heavy rainfall of 45.0–233.3 mm is observed during October followed by June and August which influences the soil population of *H. dihystera* and *H. seinhorsti*. These species were found to be maximum in number during these months, i.e., 35.2 and 14.6 per 200 g soil. During the month of October, the population of *H. dihyslera* and *H. seinhorsti* was found to be 8.79 per 5 g root and 7.3 per 5 g of root, respectively (Sivaprakash et al. 2009b).

Pathogenicity of the *Hoplolaimus seinhorsti* on red sanders was assessed by conducting an experiment in a glass house in 25 × 15 cm bags filled with steam-sterilized fine sand mixed with red earth in the ratio of 3:1. Eight days of pre-sowing treatment was done for the freshly collected seeds by presoaking continuously in normal water by changing water daily. After 15 days of germination, the plants were infested with adults and juveniles of *Hoplolaimus seinhorsti* at the rate of 0, 100, 200, 400, and 600 larvae per pot around the root zone. With the increasing level of initial inoculum, significant progressive reduction in length and weight of shoot and root was observed. The multiplication rate of *H. seinhorsti* was found to be 9.3–100 larvae per kg soil. When soil is infested with 100 larvae/kg, there is no reduction in chlorophyll content, whereas 600 larvae/kg soil significantly reduced the chlorophyll content in red sanders.

When the nematode population level exceeded 200/kg soil stomatal conductance, transpiration rate and net photosynthetic rate of *P. santalinus* (Sivaprakash et al. 2009b) are affected significantly. Sivaprakash et al. (2009a) studied stained sections of *P. santalinus* root revealed that *H. seinhorsti* acted not only as ectoparasite but also as semi-endoparasite and endoparasite. Most often it is observed that nematodes feed semi-endoparasitically causing cell collapse and deformation in the cortex forming cavities. Plant-parasitic nematodes like *Aphelenchoides*, *Helicotylenchus*, *Hemicriconemoides*, *Hoplolaimus*, *Longidorus*, *Paralongidorus*, *Pratylenchus*, *Xiphinema*, *Rotylenchulus*, *Rotylenchus*, *Siddiqia*, *Psilenchus*, *Trichodorus*, and *Tylenchorhynchus* have been reported to infect and cause disease in *Pterocarpus santalinus* (Kavitha et al. 2017)

10.5 Pest and Termite Infestation of Timber

The infestation of timber of *Pterocarpus santalinus* by heartwood borer was pragmatic in timber depots of Tirupati, Rajampet, Adurupalli, and Kadapa forest in Andhra Pradesh. Powder-post beetles are classified under the subfamily Lyctinae comprising a group of 70 species of wood-boring beetles. These make up the super family Bostrichoidea along with common furniture beetles, skin beetles, deathwatch beetles, spider beetles, and others. These powdered dust beetles are considered pests, and they infest deciduous trees, over time reducing the wood to powdery dust. In Andhra Pradesh from Nellore forest division, Adurupalli forest red sanders depot and Venkatagiri forest red sanders depot, powder-post beetles' infestation was detected. During growth cycle these powder-post beetles spend months or years inside the wood, feeding mainly on the starch content. The presence of these pests become apparent only when they emerge as an adult as they leave behind pinhole-sized openings which are often called "shot holes." They may also leave piles of powdery frass below. Depending on the species of the beetles, shot holes normally range in diameter from 0.79 to 3.2 mm. Cycle of generation continues if wood conditions are right as female beetles may lay their eggs and re-infest the wood. Infestation of termites, a group of eusocial insects of epifamily Termitoidae and order Blattodea, was observed in red sanders wood at Adurupalli and Venkatagiri depots, Nellore forest division, Andhra Pradesh (Soundararajan and Joshi 2012).

Postharvested red sanders wood is attacked by powder-post borers, sapwood borers, and termites when stored for a long time. To avoid this they should be checked immediately to evade total destruction of wood. By spraying and dipping with an appropriate solvent-based insecticide like 0.1–0.25 monocrotophos or 0.2% of paradichlorobenzene in kerosene oil, boron, or lindane formulation or by syringe injections into the flight holes, insect attack may be circumvented. By flooding the wood galleries with light organic solvent-based wood preservative formulations, having broad-spectrum of fungicides as well as insecticides, termites can be destroyed. By fumigating with methyl bromide or trimethyl borate or with ammonia, we can control termites though we cannot control re-infestation. We can also control the postharvest pest infestation of red sanders by drying the wood to 12% moisture content and by maintaining proper ventilation in the depot by air circulation. Contact insecticides should be sprayed frequently, and continuous monitoring of timber should be a top priority, and removal of completely infested piece/log regularly is a necessity (Soundararajan and Joshi 2012).

10.6 Fungal Infestation of Timbers

At Adurupalli and Venkatagiri depots of Nellore forest division, Andhra Pradesh, the wood decay fungus infestation was noticed in red sanders. The four fundamental requirements of the wood decay fungus are oxygen, favorable temperature, water, and nutrients. When the moisture content of the wood exceeds 20–30%, with optimal temperature of (32–90 °F), an adequate supply of oxygen, and nutrient source, wood-decaying fungal growth will be maximum. Variety of wood-decaying fungus digests the moist wood causing it to rot. Some wood-decaying fungus can also infest dead wood to cause brown rot. The fungi which not only grow on wood but also decay it are called lignicolous fungi. Lignicolous fungi consume wood either by digesting carbohydrate or by decaying lignin (Soundararajan and Joshi 2012). As water is the enemy of wood, moisture control must be an integral part of prevention of wood decay fungi. Borate can prove as an effective agent in killing the wood-decaying fungi (Soundararajan and Joshi 2012).

There are very limited number of studies regarding pathology of red sanders which leaves lot of scope for future research.

References

Anonymous (2007) Overview of forest pests Thailand. Forest health & Biosecurity working papers http://www.fao.org/forestry/1230301a808804d506b483ac88eb6a1f94735e.pdf. DOI 28/11/2018

Arunkumar AN, Joshi G (2014) *Pterocarpus santalinus* (Red Sanders) an endemic, endangered tree of India: current status, improvement and the future. J Trop For Environ 4:1–10

Jash A, Gupta SK, Debnath N, Dasghosh M (2018) Diversity of mites and insects occurring on dye-yielding and aromatic plants with two new records of mites from India. Intern J Sci Res 7 (9):5–7

Kavitha PG, Devanand PS, Kumaran K, Suresh KK (2017) Distribution of plant parasitic nematodes associated with tree crop nurseries. Intern J Plant Anim Environ Sci 7(1):94–100. https://doi.org/10.21276/Ijpaes

Luna RK (2005) Plantation trees. International Book Distributors, Dehradun, pp 608–614

Mohanan C, Ratheesh N, Nair LP, Rajesh Kumar KC (2005) Disease problems in root trainer forest nurseries in Kerala State and their management. http://www.metla.fi/julkaisut/workingpapers/2005/mwp011.htm. DOI 28/11/2018

National Medicinal Plants Board (2008) *Pterocarpus santalinus*. In: Agro techniques of selected medicinal plants. TERI press, New Delhi, p 159

Rajasekhar A, Reddy MN, Srivasuki KP (1994) Leaf spot disease of Red Sanders (*Pterocarpus santalinus*)- caused by *Cylindrocladium scoparium*. Indian For 120:1–6

Reddy MRS, Dayanand T (1983) Mycoflora associated with seed of Red sanders (*Pterocarpus santalinus* Linn.f.). Indian J For 6:322

Sankaran KV, Florence EJM, Sharma JK (1984) Two new diseases of forest tree seedlings caused by *Sclerotium rolfsii* in India. Eur J For Pathol 14:318–320

Sivaprakash M, Balasubramanian P, Prabhu S (2009a) Histopathology of *Hoplolaimus seinhorsti* on Bijasal and Red Sander. Ann Plant Prot Sci 17(1):209–211

Sivaprakash M, Balasubramanian P, Rathakrishnan P, Prabhu S (2009b) Pathogenicity of lance nematode, *Hoplolaimus seinhorsti* on Red Sanders, *Pterocarpus santalinus* tree seedlings. Indian J Nematol 39:54–58

Soundararajan V, Joshi SC (2012) Endemic possessions of Eastern Ghats: Red Sanders (*Pterocarpus santalinus* Linn. f.). Institute of Wood Science and Technology, Bangalore

Sundararaj RG, Shanbhag RR, Nagaveni HC, Vijayalakshmi G (2015) Natural durability of timbers under Indian environmental conditions an overview. Int Biodeterior Biodegrad 103:196–214

11 Wood Anatomy and Wood Property Variation in Red Sanders

E. V. Anoop, R. V. Rao, and Gayathri Mukundan

Abstract

Pterocarpus santalinus L. f., an elite tree species, popularly known as red sanders, is known for its unique scarlet red-colored wood ranked as one of the finest luxury woods in the globe. The exceptional wood quality is commercially valued based on the wavy grain impregnated with deep red-colored dye santalin. This precious tree is restricted in its distribution, and the quality of wood varied within the population. In this chapter the wood anatomy and variations are deliberated, and a comprehensive account on the same is discussed.

Keywords

Pterocarpus santalinus · Wood anatomy · Wood property variation

11.1 Introduction

Pterocarpus santalinus L. f. is a medium-sized deciduous tree belonging to the family Fabaceae, commonly known as red sanders and locally called as raktachandan. It has a highly restricted distribution in the southeastern portion of the Indian peninsula in which it is endemic. The Palakonda and Seshachalam hill ranges of Kadapa (Cuddapah)-Chittoor district of Andhra Pradesh are its principal geographic range; it extends northward to Kurnool and southward to Vellore (North Arcot) in Tamil Nadu (Rao and Purkayastha 1972).

The species is having erect bole and dense rounded crown, with an average height of 10 m and girth of about 1–1.5 m, and is distinct by its blackish brown bark resembling crocodile skin. Blaze exposes white-colored sapwood which turns red

E. V. Anoop (✉) · R. V. Rao · G. Mukundan
College of Forestry, Kerala Agricultural University, Thrissur, Kerala, India
e-mail: anoop.ev@kau.in

T. Pullaiah et al. (eds.), *Red Sanders: Silviculture and Conservation*,
https://doi.org/10.1007/978-981-13-7627-6_11

Fig. 11.1 Microscopic images of wood of red sanders. (After Richter and Dallwitz 2000, Commercial timbers: descriptions, illustrations, identification, and information retrieval. http://www.delta-intkey.com/wood/en/www/papptsan.htm with permission)

due to the exudation of a red gummy juice. Leaves are pinnately compound with three to five oval-orbicular leaflets and yellow flowers of typical papilionoid form.

Red sanders grows sporadically in the southernmost state of its distribution, viz., Kerala, both naturally and in plantations. The Kerala Forest Research Institute (KFRI) and the Kerala Forest Department (KFD) are maintaining research plots at Nilambur and Palappilly from 1983 onward for research work on this species.

Pterocarpus santalinus was classified as an endangered species in 1997 by the IUCN (Walter and Gillet 1998). It was first listed in the legislation of the Convention on International Trade in Endangered Species (CITES) under Appendix II to regulate trade in log, wood chips, and unprocessed broken materials, realizing the possibility of extinction, should unregulated trade continue unmitigated.

Red sanders in its population has been found to have two variabilities of wood nature, one with a wavy grain, which is envisaged as high-quality wood, and the straight-grained ones are of lower quality (Troup 1921; Lohidas and Dayanand 1984). Such "quality wood" is mostly exported to Japan for making a musical instrument called shamisen. Also, small carvings and furniture are made from it, and consequently, demand is higher for the figured, quality wood. The tree is renowned for its characteristic: exquisite color, beauty, and unique technical qualities (Figs. 11.1, 11.2, 11.3, and 11.4).

The red-colored santalin pigment in heartwood was earlier used to produce textile dyes; however, it is now being substituted with synthetic dyes. Heartwood extracts are used in Indian Ayurvedic medicines for the treatment of skin diseases, leprosy,

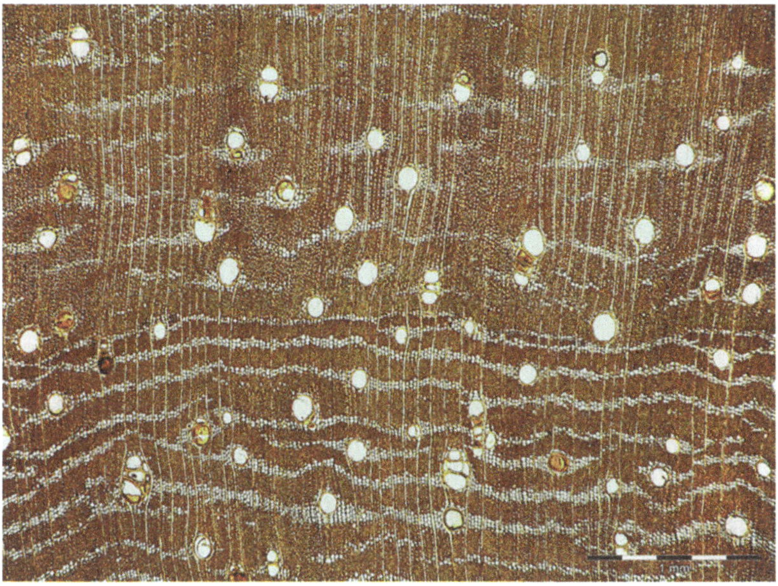

Fig. 11.2 Transverse section of wood of red sanders. (After Richter and Dallwitz 2000, Commercial timbers: descriptions, illustrations, identification, and information retrieval. http://www.delta-intkey.com/wood/en/www/papptsan.htm with permission)

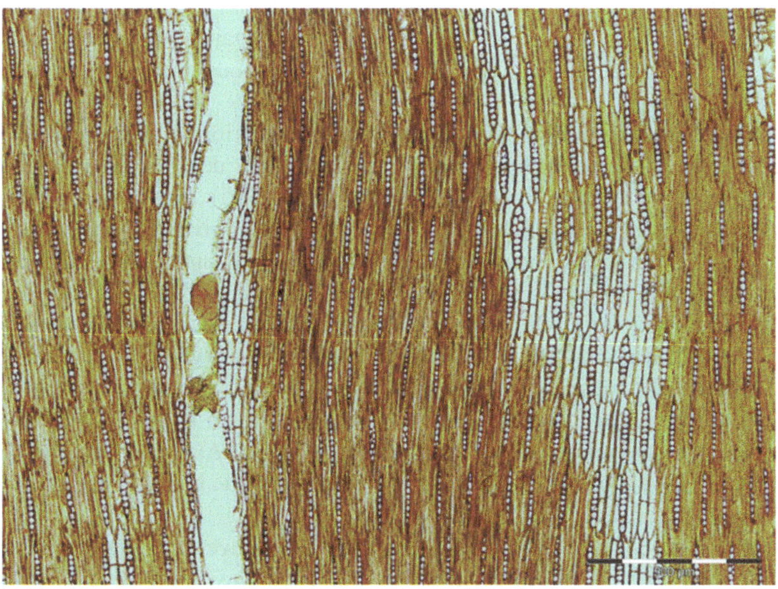

Fig. 11.3 Tangential section of wood of red sanders. (After Richter and Dallwitz 2000, Commercial timbers: descriptions, illustrations, identification, and information retrieval. http://www.delta-intkey.com/wood/en/www/papptsan.htm with permission)

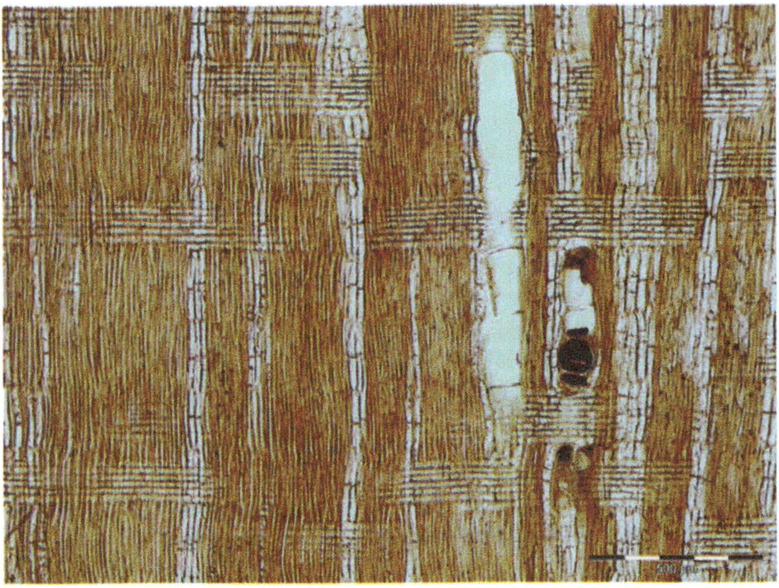

Fig. 11.4 Radial section of wood of red sanders. (After Richter and Dallwitz 2000, Commercial timbers: descriptions, illustrations, identification, and information retrieval. http://www.delta-intkey.com/wood/en/www/papptsan.htm with permission)

swellings, ulcers, fever, diarrhea, diabetics, and even against poisoning (Gupta and Uniyal 2003). Santalin also finds its use in coloring food stuff and pharmaceutical preparations.

Despite its protection as a "reserved species" made prior to independence (Troup 1921), red sanders was overexploited in the following decades as well up to the 1960s in search of quality timber for export (Green 1995). Furthermore, protection was assigned under the Andhra Pradesh preservation of private forest rules (1978) and felling ban in state forests since 1982 (Mulliken and Crofton 2008). The smuggling of red sanders wood from Andhra Pradesh and Tamil Nadu is more than those reported from other states, indicating the high quality of wood grown in these states. It was noticed that smuggled red sanders was mainly used in the manufacture of the musical instrument shamisen, as stated earlier. However, newspaper reports revealed that smuggling to China and Japan is mainly for the use in nuclear reactors and related experiments (Rao, M.M, 3rd August 2012, The Hindu). However, detailed information, particularly from scientific literature is not available yet (Bhagyaraj and Ramana 2013).

Since the uniqueness of wood quality from certain provenance regions is reported and thus raised several research questions, a study was initiated to interpret the physical and anatomical properties of red sanders in Kerala and to compare the wood quality variation in provenances from certain selected regions in Andhra Pradesh and Tamil Nadu (Figs. 11.5, 11.6, 11.7, 11.8, 11.9, 11.10, 11.11, and 11.12).

Fig. 11.5 Cross-sectional disc from Kadapa, AP

Fig. 11.6 Heartwood measurements using the software Digimizer, version 2.0

11.2 Distribution of Red Sanders

Pterocarpus species (*Pterocarpus santalinus*, *Pterocarpus dalbergioides*, *Pterocarpus indicus*) are widely distributed in tropics throughout the world, especially in India, Sri Lanka, Taiwan, and China. Naturally, red sanders is found in Kadapa (earlier known as Cuddapah), Kurnool-Chittoor, and Nellore districts of Andhra Pradesh and a few adjoining areas of Tamil Nadu. In Tamil Nadu it occurs rarely up to a 500 m elevation. Physiographically 300–800 m is the most favorable altitude (Subasinghe and Yoon 2011), and it was also found to be growing on hilly terrain and slopes occupied by very shallow to shallow-brown sandy loam soil. In its natural habitat, it grows well in temperature from 13 to 37 °C and rainfall from 35 to 135 cm.

Fig. 11.7 Wood specimens used for physical property analysis and anatomical studies

Fig. 11.8 Powdered wood used for heartwood color estimation by using Munsell color chart

11.3 Wood Properties

11.3.1 Physical Properties

11.3.1.1 Specific Gravity

Specific gravity can be considered as a reliable index of wood quality which reflect the amount of wood substance deposited per unit volume of tree trunk and thus is a factor influencing the amount of forest biomass (Wiemann and Williamson 2013). The specific gravity of wood showed considerable difference between species within

Fig. 11.9 (**a**) Cross section of Thrissur sample TS 4X. (**b**) Microtome section of Thrissur sample TS 10

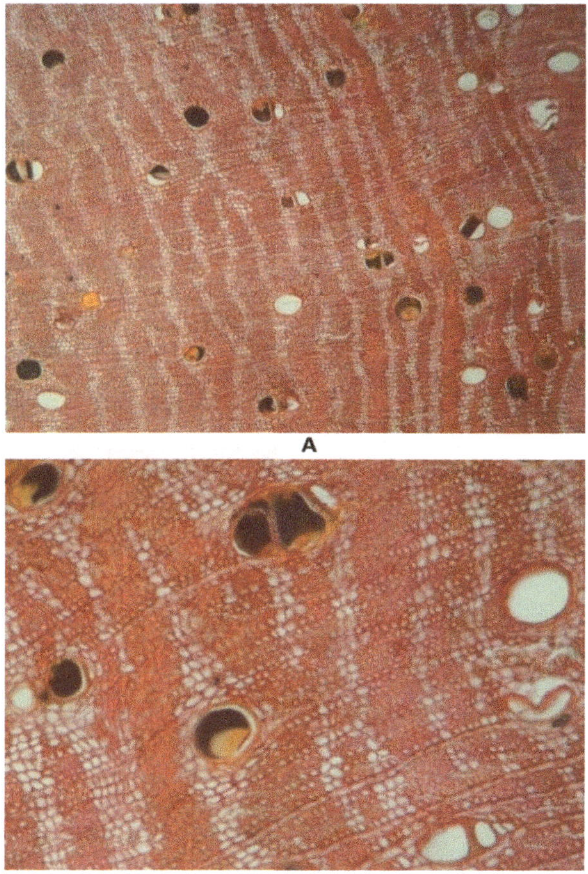

A

B

each region and between regions. Negative correlation between density and mean growth rate per species has been observed in several tropical forests (Landau 2004). Wood specific gravity is one of the important factors that determine the economic utility of wood. In hardwood species, stiffness and strength are strongly influenced by wood density and microfibril angle, the proportion of lignin, extractives and interlocked grain, and the extent of spiral grain.

11.3.1.2 Shrinkage and Swelling

Shrinkage is one of the factors that reflect the anisotropic property of wood. Generally, wood shows a very complicated swelling or shrinkage behavior with the adsorption or desorption of bound water below fiber saturation point. The increase or decrease of free water above fiber saturation point does not cause changes in wood dimensions. The swelling or shrinkage of wood shows anisotropy, and the ratio is generally 10 (tangential): 5 (radial): 0.1–1 (longitudinal). Considerable

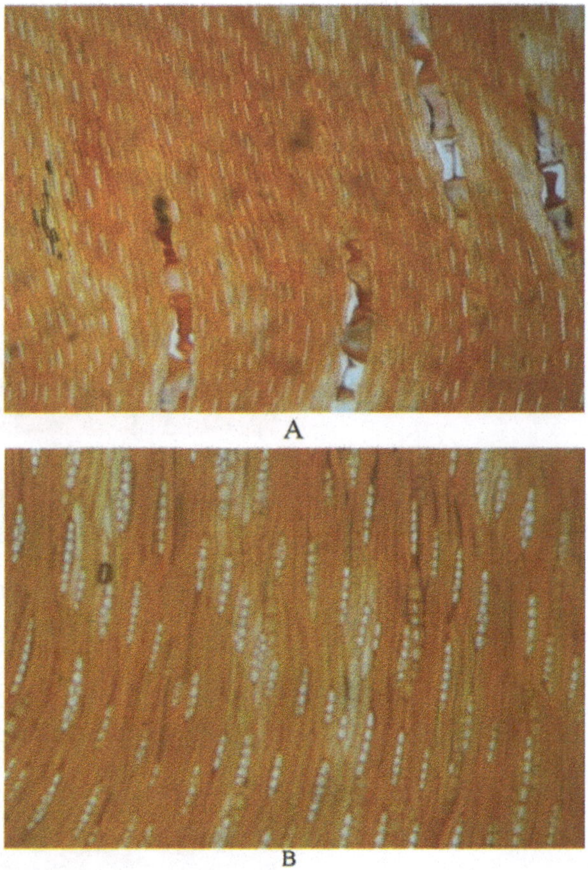

Fig. 11.10 (**a**) Tangential section of Thrissur sample TLS 4X. (**b**) Tangential section of Thrissur sample TLS 10X

dimensional changes that occur due to swelling or shrinkage are the cause of cracks in lumber and internal stresses and are undesirable characteristics from the point of view of timber utilization. On the other hand, swelling and shrinkage could also be considered as intelligent characteristics of wood because wood can change dimensions by itself in response to the atmosphere (Okuma 1998). Therefore, it is important to obtain accurate knowledge on the swelling and shrinkage behavior of wood. Generally wood shows a very complicated swelling or shrinkage behavior with the adsorption or desorption of bound water below fiber saturation point. The increase or decrease of free water above fiber saturation point does not cause changes in wood dimensions. It was also observed that the relationship between longitudinal shrinkage and microfibril angle was different in compression wood compared to normal wood, with shrinkage in compression wood being much more sensitive to changes in microfibril angle. Compression wood and spiral grain are also more prevalent in juvenile wood than in mature wood and contribute to excessive longitudinal shrinkage. Furthermore, longitudinal shrinkage is greater in early juvenile wood than in late juvenile wood.

Fig. 11.11 (**a**) Radial section of Thrissur sample RLS 4X. (**b**) Radial section of Thrissur sample RLS 10X

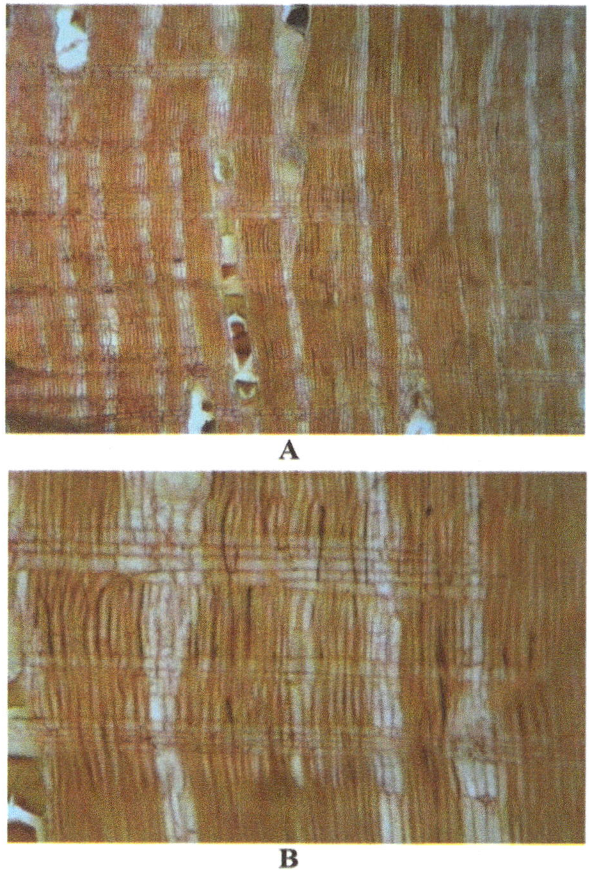

Shukla et al. (1999) carried out studies on shrinkage properties in plantation grown Andaman padauk (*Pterocarpus dalbergioides*) and compared those values with the natural grown Andaman padauk and that of teak. The results on average radial (R), tangential (T), longitudinal (L), and volumetric shrinkage from green to oven-dry condition indicated that the Andaman padauk had all those values comparable to that of teak.

11.3.1.3 Moisture Content

Moisture content in wood is measured by finding the amount of water present in a piece of wood relative to wood itself. It has direct effect on weight, strength, physical behavior, and processing characteristics that influences its utilization. The amount of water present in each piece of wood is expressed as the percentage of weight of the water to the oven-dry weight (Simpson and Tenwolde 1999). Moisture content in trees can range from 30% to more than 200% of weight of the wood substances.

Moisture content and specific gravity are negatively related within a species; the higher the specific gravity, the lower the moisture content. It is found that moisture

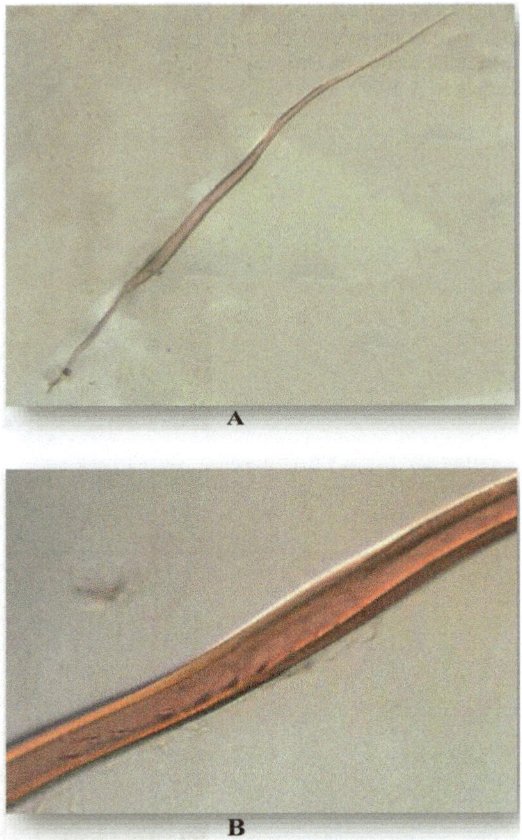

Fig. 11.12 (**a**) Macerated section showing fiber in Thrissur sample 10X. (**b**) Macerated section showing fiber in Thrissur sample 40X

content differs significantly with respect to species. Different species and different specimens of the same species show varied responses to percentage change in strength, due to given changes in moisture content. In general, strength properties increase with decrease in moisture content for clear wood. Mechanical properties of wood alter with moisture content (Gerhards 1982). The mechanical properties of wood increase as moisture content decrease below fiber saturation point, at least down to about 5% MC.

11.3.2 Anatomical Properties

Anatomical properties of wood can be influenced to different degrees by environmental and genetic factors. Variation of anatomical properties of wood within the main stem of a tree is often quite large and has been shown to be strongly related to proximity of crown during wood formation (Larson 1962, 1964). In general, all anatomical properties will vary significantly from pith outward within a tree. Mature wood and juvenile wood can be distinguished by anatomical properties.

11.3.2.1 Vessel morphology

Vessel morphology is substantially influenced by environmental factors, genetic engineering, silviculture, and treatment with plant growth substances. Fichtler and Worbes (2012) studied the variation in wood anatomical variables within and between families and sites. Within all groups, both families and sites, we could observe a high variability of vessel diameter. Basically, an increase of variation with increasing number of individuals within a group was observed. Within the category "plant family," they observed the highest variation of vessel diameter within the pantropical family Fabaceae, a group with a high number of species and individuals. Lauraceae and Myrtaceae showed the lowest variation in vessel diameter.

Vessel diameter and vessel element length decrease with increase in aridity and that of vessel numbers increase. Vessel lumen diameter generally shows a juvenile to mature pattern of variation from pith to the bark (radial variation), where vessel lumen diameter is smaller in the inner part of stem, and gradually increases in size outward before leveling off in the outer part of the stem.

11.3.2.2 Ray Morphology

Ray morphology has significant contribution to various wood properties. Variation in wood structure associated with difference in growth rate concerns the development of rays. In fast-grown timber, the rays tend to become more numerous, wider, or higher, than in slow-grown wood; there are marked increases in the total number of the ray cells per unit tangential area of the wood. Varying differences of this nature have been described in hardwoods (White and Robards 1996).

11.3.2.3 Wood Fiber Morphology

Fibers are narrow elongated cells with tapering end that give mechanical support to wood. It is one of the significant features that influence the strength and shrinkage properties.

The external factor plays a key role in bringing about physiological changes in trees there by affecting the cambial activity. The tree to tree variation in wood properties within a species and also within the plantation are large which can be observed throughout the variation in physical and anatomical properties. The fast-grown timber showed higher numerical values for anatomical properties, viz., fiber length, fiber diameter, fiber lumen diameter, and double wall thickness of fiber when compared to slow-grown timber. Interestingly the intra-tree variation revealed that the outer region of wood in both types of trees had longer and wider vessels. Though variation between trees and among trees is nonsignificant, numerically superior values are recorded in fast-grown tree than slow-grown tree (Swaminathan et al. 2012).

The mean value of lumen diameter appeared to be larger in the wavy-grained trees of red sanders as compared to that from normal-grained trees, and the difference between the mean values was found to be very highly significant. This is likely to be a useful character for screening the saplings if it is confirmed after an examination of a larger number of trees (Kedharnath and Rawat 1976).

11.3.2.4 Provenance Variation

Over immense time variation occurs among segments of a species that grow in different environment within a species ranges due to evolutionary selection (Callaham 1964). These species with different characteristics represent provenances or geographic race. It has defined geographic race as a subdivision of a species consisting of genetically similar individuals, related by common descent, and occupying a territory to which it has become adapted through natural selection.

Species with an extensive geographic distribution such as teak, which extend from Kerala in the south to Arunachal Pradesh in the north, contain much more genetic diversity than species with very limited distribution such as red sanders which has restrictive distribution to small part of Andhra Pradesh and Tamil Nadu.

The feasibility of utilizing morphological markers for determining existing provenance variation in the African savanna tree *Pterocarpus angolensis* was assessed. The study aimed to investigate the presence and extent of variability among and within *P. angolensis* populations from five provenances and to assess the utility of morphological markers in determining provenance variation in *P. angolensis*. Result showed the difference in height and growth between provenances of *P. angolensis* within 1 month of germination, but the variation was not related to the geographic distance between sources (Chisha-Kasumu et al. 2009).

11.3.3 Effect of Provenance Variation in Wood Quality

Extensive experience has shown that the only real way to determine what the wood will be like when trees of a given provenance are grown in a new environment is through growing that tree directly there. Environmental causes for these differences are similar to factors that affect growth rate, such as elevation, latitude, soils, rainfall, and others. In addition to this, bud flushing and growth cessation characteristics are the two specific growth characteristics that have been reported to be closely related to provenance and wood qualities.

11.3.4 Different Sources of Wood Variation

Wood is a variable substance with difference occurring among species and genera, among geographical sources within a species and among trees within geographic sources as well as within each individual tree. Tree-to-tree variability is especially large, with difference within a species often under strong genetic influences. External factors such as environment, site conditions, and silviculture treatment also have impact on regular pattern of wood variation, and these are regarded as extrinsic.

Variation in climate and soil results in inter-site differences in the tree life history strategies within a community, which has important implications for ecosystem structure and dynamics. Mean wood specific gravity (oven-dry weight divided by fresh volume, sometimes also referred to as wood density in the literature) differed significantly among sites, varying inversely with soil fertility and independent of rainfall, seasonality, and temperature. Within sites wood specific gravity varied widely among species. In some places species variation was significantly, weakly, negatively correlated with sapling and tree mortality and relative growth rates. Altogether, the results suggest that the distribution of tree life history strategies in a community varies substantially among sites, with important consequences for community and ecosystem properties such as aboveground carbon stores (Landau 2004).

The inherent variability is one of the important ones and has consequences of a very different kind. Even when much of the natural variability (such as that due to knots, cracks, and irregularities of grain) has been excluded, as in laboratory tests of small clear specimens, there remains a considerable residue of variation; this is evident, for instance, in strength properties of timbers.

11.3.5 Dendroprovenancing

Illegal logging is a major cause of worldwide deforestation, and demand for scientific methods to identify the origin of timbers is increasing. Dendroprovenancing is one such method which reveals the origin of unknown wood by calculating correlation of ring-width series of unknown reference tree of known geographic origin.

Archeological studies with the help of dendroprovenancing and dendrochronology have helped archeologist to get an idea on timbers' import, trade, and trace backing of the timber found in many Western and European countries to their respective colonies.

11.4 Wood Anatomy of Red Sanders

Wood anatomy of red sanders has been studied by Richter and Dallwitz (2000), Vijendra Rao and Chauhan (2003), MacLachlan (2009), Anonymous (2010), Arun Kumar et al. (2017), and Suresh et al. (2017), and the details of the wood anatomy are given below.

Heartwood ranges from a dark orange to a deeper reddish purple, often with darker streaks throughout. Colors tend to darken significantly over time to deep reddish purple to nearly black, without streaks. Pale white sapwood is narrow and is clearly demarcated from the heartwood. Sometimes seen with a wavy, interlocked grain figure with a density (0.58–) 0.77–0.85 g/cm^3. Grain is generally straight or slightly interlocked. With a medium uniform texture and high natural luster. Wood is diffuse-porous. Vessels arranged in no specific pattern, in multiples, commonly in short (2–3 vessels) radial rows. Two distinct vessel diameter classes absent. Average tangential vessel diameter 115–230–295 μm. Average number of vessels/mm^2 1–2. Perforation plates simple. Intervessel pits alternate, average diameter (vertical) (8–) 9–12 (–13) μm, pits vestured. Vessel-ray pits with distinct borders or with reduced borders or apparently simple, similar to intervessel pits or different from intervessel pits, rounded or angular. Helical thickenings absent. Other deposits present (orange brown). Fibers of medium wall thickness to very thick-walled. Average fiber length 1035–1380–1660 μm. Fiber pits mainly restricted to radial walls, simple to minutely bordered. Fibers nonseptate. Axial parenchyma banded or not banded. Bands marginal (or seemingly marginal). Bands fine. Axial parenchyma apotracheal, or paratracheal (short tangential lines without contact to vessels). Apotracheal axial parenchyma diffuse and diffuse-in-aggregates. Paratracheal axial parenchyma aliform and confluent. Aliform parenchyma of the winged type. Axial parenchyma fusiform and as strands. Average number of cells per strand: 2. Rays 12–17 per tangential mm, exclusively uniseriate or multiseriate (in some specimens also with few biseriate rays). Height of large rays up to 500 μm. Rays composed of a single cell type (homocellular); homocellular ray cells procumbent. Storied structure present, all rays storied, axial parenchyma storied, vessel elements storied, fibers storied. Number of ray tiers per axial millimeter 4–6. Crystals present, prismatic, located in axial parenchyma cells. Crystal-containing axial parenchyma cells chambered. Number of crystals per cell or chamber one. Silica not observed.

11.4.1 Chemical Testing of Redsanders

Heartwood not fluorescent. Water extract fluorescent (light greenish blue); water extract colorless to brown (light orange brown). Ethanol extract fluorescent (rosé to orange, of variable intensity). Ethanol extract colorless to brown or red. Froth test positive. Splinter burns to partial ash. Ash white to gray.

11.5 Wood Property Variation in Red Sanders

11.5.1 Physical and Anatomical Properties

We carried out studies on variation in physical and anatomical properties of red sanders (*Pterocarpus santalinus*) between provenances of Thrissur (Kerala) with Thiruvallur (Tamil Nadu) and Kadapa (Andhra Pradesh) regions, and the results are given below.

Various parameters of vessels, fibers, and rays including their length, diameter, and number per unit area were noted from the image using the computer software (Labomed Digi Pro-2). Each of the above measurements was taken ten times from each sample and was expressed in micrometers (μm).

From the tangential section of the wood samples, parameters such as ray height, ray width, and ray frequency were measured. Ray frequency was measured by counting the number of rays from a selected area in the section with the help of the image analysis software Labomed Digi 2 (Fig. 11.6) and was expressed in number per millimeters (mm) (Anoop et al. 2005).

Maceration of wood samples was done using Jeffrey's method. Jeffrey's solution was prepared by mixing equal volumes of 10% potassium dichromate and 10% nitric acid. Wood shavings were taken from the 1 cm^3 wood blocks from the pith, middle, and periphery portion of the wood blocks. These chips were boiled in Jeffrey's solution for 10–15 min until the fiber becomes separated. After that, the test tubes were kept intact for 5–10 min, so that the fibers settled in the bottom of the tube. The surface solution was discarded, and the remaining fiber in the tube was washed thoroughly with distilled water. This process was repeated until the traces of acid solution were removed. Separated fibers were stained using safranin and mounted on temporary slides using glycerine as the mountant.

From the temporary slides prepared, parameters such as fiber diameter, fiber wall thickness, and fiber lumen diameter were measured using Labomed Digi 2.

$$\text{Runkel ratio} = \frac{2 \times \text{Fiber wall Thickness (FWT)}}{\text{Fiber Lumen Diameter (FLD)}}$$

$$\text{Slenderness ratio} = \frac{2 \times \text{Fiber length(FL)}}{\text{Fiber Diameter (FD)}}$$

$$\text{Rigidity coefficient} = \frac{2 \times \text{Fiber wall Thickness (FWT)}}{\text{Fiber Diameter (FD)}}$$

$$\text{Flexibility coefficient} = \frac{\text{Fiber Lumen Diameter (FLD)}}{\text{Fiber Diameter (FD)}}$$

$$\text{Shape factor} = \frac{D^2 - L^2}{D^2 + L^2}$$

D – Fiber width
L – Lumen width

$$\text{Vessel vulnerability} = \frac{\text{Vessel diameter}}{\text{No of vessels per sq.mm}}$$

The dimensions of the samples were determined using the software Digimizer.

One sample t-test was done to find the variation in physical and anatomical properties of red sanders (*Pterocarpus santalinus*) between provenances of Thrissur (Kerala) with Thiruvallur (Tamil Nadu) and Kadapa (Andhra Pradesh) regions. Results have been listed under the headings: (1) Physical Properties and (2) Anatomical Properties.

11.6 Physical Properties

11.6.1 Moisture Content

For physical property analysis, moisture content of the sample at air-dry, green, and oven-dry conditions was studied. There was significant variation observed in green condition between Thiruvallur and Thrissur samples (5% level). Among the three samples collected, higher value of moisture content was found for Thiruvallur (33.23%). At air-dry condition and green condition, moisture content of 16.458% and 26.08% was obtained, respectively, for Thrissur sample.

At air-dry condition, no significant difference between samples of Kadapa and Thrissur was observed. Lower value of moisture percentage (10.53%) in air-dry condition was observed for Thiruvallur sample. The Thrissur sample had showed high moisture content in air-dry condition (26.08%) and Kadapa sample with a lower moisture content (16.2%) among all the samples at green condition.

11.6.2 Specific Gravity

Specific gravity in air-dry condition shows significant variation (5%) between samples of Thiruvallur (1.04) and Kadapa (1.2) with Thrissur (0.98). The values of specific gravity at oven-dry condition of Thiruvallur (0.97) and Kadapa (1.14) showed significant variation at 5% level. The values of specific gravity at green condition for Thiruvallur (1.245) and Thrissur (0.96) sample were found, and no significant variation (at 5%) was seen in green and oven-dry for two samples with Thrissur.

11.6.3 Shrinkage

11.6.3.1 Radial Shrinkage

Radial shrinkage of the samples at three conditions (air-dry, green, and oven-dry) was studied. At green to air-dry condition, 5% significance was observed for Thrissur sample with Kadapa and Thiruvallur. All three conditions were significant at 5% level for sample between Thrissur and Kadapa. Values of shrinkage for sample from Thrissur were higher in all the three conditions in comparison with the other two samples collected.

11.6.3.2 Tangential Shrinkage

At various conditions three samples were studied for tangential shrinkage. The variation from Thrissur to Kadapa and to Thiruvallur observed 5% significance for the test conducted for all the three conditions (green to air-dry, green to oven-dry, and air- to oven-dry). No significant difference was found for shrinkage between Thiruvallur and Kadapa samples. Tangential shrinkage (green to oven-dry) was found to be significant at 1% level.

The tangential shrinkage of green to oven-dry condition of Thrissur sample was observed to be 1.95% and for Thiruvallur 0.35%, while for sample from Kadapa, it was 0.30%. Considering the tangential shrinkage of green to air-dry condition, the values of Thrissur, Thiruvallur, and Kadapa sample were found to be 0.58%, 0.35%, and 0.30%, respectively.

11.6.4 Coefficient of Anisotropy

Coefficient of anisotropy (ratio of tangential to radial shrinkage) of the wood samples was studied (Fig. 11.10). Values for air-dry to oven-dry, green to oven-dry, and green to air-dry were observed to be significant at 5% level between samples of Thrissur with Thiruvallur and Kadapa. Coefficient of anisotropy (green to air-dry) of Thrissur sample was 0.9, Thiruvallur sample was 1.2, and that of Kadapa was 1.25. At air-dry to oven-dry condition, the value of coefficient of anisotropy was 1.5 for Thrissur, 0.82 for Thiruvallur, and 0.86 for Kadapa. While considering green to oven-dry condition, the value was 1.03 for both samples from Thiruvallur and Kadapa and 1.3 for Thrissur sample.

11.6.5 Heartwood Color

The color variations in heartwood were studied with the help of Munsell color chart. Heartwood color-determining components such as hue, chroma, and value (Table 11.1) did not show any significant variation among the samples. The heartwood color of Kadapa, Thiruvallur, and Thrissur was observed as red.

Table 11.1 Variation in heartwood color components among Thrissur, Thiruvallur, and Kadapa samples

Parameters		Sample	Value
Heartwood color	Hue	Thrissur	2.5 YR
		Thiruvallur	2.5 YR
		Kadapa	2.5 YR
	Value	Thrissur	4
		Thiruvallur	4
		Kadapa	4
	Chroma	Thrissur	8
		Thiruvallur	6
		Kadapa	8

11.7 Anatomical Properties

11.7.1 Vessel Morphology (Figs. 11.9, 11.10, and 11.11)

Vessel diameter of Thrissur sample was 160.11 μm. The results of vessel diameter were found to be significant between the samples of Thrissur and Kadapa. Highest vessel diameter was shown by Thiruvallur sample with a value of 434.78 μm, and for Kadapa sample, it was 158.75 μm.

Vessel area was nonsignificant for samples with Thrissur. It was found to have significant variation between the Thiruvallur and Kadapa samples. Rationally, a greater value of vessel area was observed in Thiruvallur sample (22,440.08 μm^2) than that of Kadapa (19,915.48 μm^2) and Thrissur samples (26,710.647 μm^2).

The results for vessel frequency revealed that there was no significant difference between the three samples. The vessel frequency value for the studied samples ranged from 1 to 3 vessels per mm^2. The lower value was observed for the Kadapa sample (2 vessels per mm^2) and was found to have not much significant difference with the Thiruvallur sample (3 vessels per mm^2).

Vessel vulnerability value claims no significant variation with regard to the three sample locations. Vessel vulnerability value was observed to be lower in the sample collected from Thrissur (51.42) than Thiruvallur (174.61) and Kadapa (105.83).

11.7.2 Ray Morphology (Figs. 11.9, 11.10, and 11.11)

All parameters of ray such as ray height, ray width, and ray frequency were studied. Ray height was found to be significant (5%) for Thrissur and Thiruvallur samples. Thrissur sample showed no significance (5%) for ray height (168.68 μm) and ray width (16.5 μm) when compared with Thiruvallur and Kadapa samples. Ray height values of Thiruvallur and Kadapa samples were found to be 187.32 μm and 189.5 μm, respectively.

Ray width value was higher in Thiruvallur (22.85 μm) than Thrissur (16.51 μm) and Kadapa (19.5 μm). Ray height was found to be highly significant at 1%, whereas ray width and ray frequency were significant at 5%. An average of 35 rays per mm^2 was observed in Thiruvallur sample, and for Kadapa it was 32 rays per mm^2.

11.7.3 Fiber Morphology (Figs. 11.9, 11.10, and 11.11)

Fiber morphological features such as fiber length, fiber lumen width, fiber wall thickness, and fiber width as well as fiber ratios such as Runkel ratio, slenderness ratio, flexibility coefficient, rigidity coefficient, and shape factor were also studied. The variation was significant for fiber length (b/w Thrissur and Kadapa) and fiber width (b/w Thrissur and Thiruvallur) at 5% level. Sample from Thrissur showed a fiber length of 773.5 μm and fiber width of 26.27 μm, respectively.

While comparing the values of fiber wall thickness, Thrissur sample has a least difference in fiber wall thickness (6.275 μm) with Thiruvallur (7.9 μm) and Kadapa samples (7.3 μm). Fiber length was more for Thiruvallur sample (1048.9 μm) in comparison with Thrissur (773.5 μm) and Kadapa (840.5 μm).

Fiber ratios such as Runkel ratio, flexibility coefficient, rigidity coefficient, and shape factor were observed nonsignificant between the three samples. Slender ratio (61.97) of Thrissur sample is significant (5%) over the sample from Kadapa (70.86). Flexibility coefficient (54.183) and rigidity coefficient (47.81) for Thrissur sample are comparatively less with Thiruvallur and Kadapa samples. Shape factor is more for sample from Kadapa (0.76) and least having 0.53 for Thrissur.

Salient results of the study are summarized in Tables 11.2 and 11.3.

Table 11.2 Comparison of physical and anatomical properties between samples obtained from Thiruvallur and Thrissur

Sl. No	Properties	Mean	SD	T value
1	Specific gravity (air-dry)	0.99*	0.23	−0.56
2	Specific gravity (wet)	0.97	0.09	−8.05
3	Specific gravity (oven-dry)	0.89	0.08	−4.35
4	Fiber length (μm)	773.51	262.01	−4.70
5	Fiber width (μm)	26.28*	4.98	0.65
6	Fiber wall thickness (μm)	6.28	1.62	−4.53
7	Fiber lumen width (μm)	14.22	4.26	3.34
8	Runkel ratio	0.97	0.43	−5.17
9	Slenderness ratio	61.40	27.39	−3.40
10	Flexibility coefficient	54.70	13.69	−10.24
11	Rigidity coefficient	48.87	12.80	−4.74
12	Shape factor	0.54	0.17	−3.70
13	MC at wet weight basis (%)	26.09*	9.14	−1.75
14	MC at dry weight basis (%)	16.46	1.64	8.07
15	Ray height (μm)	168.69	34.14	−2.44
16	Ray width (μm)	16.51	5.67	−5.00
17	Ray frequency (No/mm^2)	36.77*	5.98	1.70
18	Vessel area (μm^2)	26,710.65	8616.49	2.28
19	Vessel frequency (No/mm^2)	160.11	29.89	23.58
20	Vessel diameter (μm)	160.11	29.89	−41.10
21	Vessel vulnerability	51.42	21.75	−25.34
22	Radial shrinkage (Grn to AD)	0.62*	0.43	1.72
23	Radial shrinkage (AD to OD)	0.86	0.92	1.51
24	Radial shrinkage (Grn to OD)	1.49	0.79	2.69
25	Tangential shrinkage (Grn to AD)	0.59*	0.40	1.34
26	Tangential shrinkage (AD to OD)	1.35*	1.47	1.75
27	Tangential shrinkage (Grn to OD)	1.95*	1.71	1.83
28	Coefficient of anisotropy (Grn to AD)	1.35*	0.83	0.39
29	Coefficient of anisotropy (AD to OD)	3.75*	3.52	1.86
30	Coefficient of anisotropy (Grn to OD)	2.02*	2.30	0.96

*Significant at 0.05 level

Table 11.3 Comparison of physical and anatomical properties between samples obtained from Kadapa and Thrissur

Sl. No	Properties	Mean	SD	T value
1	Specific gravity (air-dry)	0.99*	0.23	−2.40
2	Specific gravity (wet)	0.96	0.079	−8.20
3	Specific gravity (oven-dry)	0.81	0.084	−8.85
4	Fiber length (μm)	773.50*	262.01	−1.14
5	Fiber width (μm)	26.28	4.98	2.11
6	Fiber wall thickness (μm)	6.28	1.62	−2.82
7	Fiber lumen width (μm)	14.22	4.26	6.17
8	Runkel ratio	0.97	0.43	−7.56
9	Slenderness ratio	61.98*	27.39	−1.55
10	Flexibility coefficient	54.70	13.69	−5.96
11	Rigidity coefficient	48.87	12.80	−4.67
12	Shape factor	0.54	0.17	−5.80
13	MC at wet weight basis (%)	26.09*	9.14	2.44
14	MC at dry weight basis (%)	16.46*	1.64	5.04
15	Ray height (μm)	168.69	34.15	−2.73
16	Ray width (μm)	16.51	5.67	−2.36
17	Ray frequency (No/mm^2)	36.77	5.98	3.34
18	Vessel area (μm^2)	26,710.65	8616.49	3.53
19	Vessel frequency (No/mm^2)	160.11	29.89	23.73
20	Vessel diameter (μm)	160.11	29.89	0.20
21	Vessel vulnerability	51.42	21.75	−11.19
22	Radial shrinkage (Grn to AD)	0.62*	0.43	1.99
23	Radial shrinkage (AD to OD)	0.86*	0.92	1.41
23.24	Radial shrinkage (Grn to OD)	1.49*	0.79	2.73
25	Tangential shrinkage (Grn to AD)	0.59*	0.40	1.69
26	Tangential shrinkage (AD to OD)	1.35*	1.47	1.69
27	Tangential shrinkage (Grn to OD)	1.95*	1.72	1.85
28	Coefficient of anisotropy (Grn to AD)	1.34*	0.83	0.25
29	Coefficient of anisotropy (AD to OD)	3.75*	3.52	1.86
30	Coefficient of anisotropy (Grn to OD)	2.02*	2.30	0.96

*Significant at 0.05 levels

11.8 Conclusion

- Specific gravity at green condition for Thiruvallur and Thrissur samples has no significance (at 5% level) for green and oven-dry condition for the two samples with Thrissur sample.
- Significant variations observed in green condition between Thiruvallur and Thrissur samples (5% level).
- For all the three conditions studied (green to air-dry, green to oven-dry, and air- to oven-dry), variation from Thrissur to Kadapa and to Thiruvallur was observed at

5% significance. No significant difference was found for shrinkage between Thiruvallur and Kadapa samples.

- At all the three conditions, radial shrinkage showed significant difference for sample between Thrissur and Kadapa.
- Radial shrinkage for green to air- and oven-dry condition was found to be significant between samples obtained from Thiruvallur and Kadapa.
- The values of coefficient of anisotropy for air-dry to oven-dry, green to oven-dry, and green to air-dry were observed to be significant at 5% level between samples of Thrissur with Thiruvallur and Kadapa.
- Coefficient of anisotropy (green to air-dry) is considerably significant at 1% level between Thiruvallur and Kadapa samples.
- The heartwood color of Thrissur, Thiruvallur, and Kadapa samples is observed to be red in color which on exposure turns dark brown.
- The difference in vessel diameter value was only found significant for sample from Thrissur with Thiruvallur and Kadapa.
- Significant variation was observed for vessel area between samples from Thiruvallur and Kadapa.
- Vessel frequency and vessel vulnerability claim no significant variation for Thrissur sample with Kadapa and Thiruvallur.
- Only ray frequency was found to show significant variation for the sample from Thrissur, compared with Kadapa and Thiruvallur.
- But all ray morphological parameters were significantly varying for the sample from Thiruvallur and Kadapa.
- The variation was significant for fiber length between samples of Thrissur and Kadapa and for fiber width between Thrissur and Thiruvallur.
- Fiber ratios such as Runkel ratio, flexibility coefficient, rigidity coefficient, and shape factor were observed nonsignificant between the three samples.
- Slender ratio of Thrissur sample is significant (5%) with sample from Kadapa.

References

Anonymous (2010) *Pterocarpus santalinus* L. anatomical properties. http://iwst.icfre.gov.in/database/Redsanders/properties/anatomical.htm

Anoop EV, Antony F, Bhat KV, Davis LA, Babu LC (2005) Anatomical key for the identification of important Timbers of Kerala. Kerala Agricultural University Kerala State Council for Science Technology and Environment, Thiruvananthapuram, pp 39–41

Arun Kumar AN, Joshi G, Manikandan S (2017) Variability for heartwood content in three commercially important tree species of Peninsular India—*Hardwickia binata, Pterocarpus santalinus* and *Santalum album*. In: Pandey K, Ramakantha V, Chauhan S, Arun Kumar A (eds) Wood is good. Springer, Singapore, pp 117–126. https://doi.org/10.1007/978-981-10-3115-1_12

Bhagyaraj A, Ramana DV (2013) Status of Red Sanders in business. Indian J Appl Res 3 (7):2249–2555

Callaham RZ (1964) Provenance research: investigation of genetic diversity associated with geography. Unasylva 18:40–50

Chisha-Kasumu E, Woodward S, Price A (2009) Phenotypic variation among five provenances of *Pterocarpus angolensis* in Zimbabwe and Zambia Southern forests. A J Forest Sci 71(1):41–47
Fichtler E, Worbes M (2012) Wood anatomical variables in tropical trees and their relation to site conditions and individual tree morphology. Int Assoc Wood Anat J 33(2):625–633
Gerhards CC (1982) Effects of moisture content and temperature on the mechanical properties of wood: an analysis of immediate effects. J Wood Fiber Sci 14(1):4–36
Green CL (1995) Non-wood products: natural colorants and dye stuff. FAO, Rome
Gupta S, Uniyal BM (2003) Indian woods–their medicinal importance and identification. Indian For 129:1225–1240
Kedharnath S, Rawat MS (1976) Studies on variation in fiber morphology in wavy-grained and straight grained trees of Red Sanders. Indian For 102(7):132–143
Landau MHC (2004) Interspecific and intersite variation in wood specific gravity of tropical trees. Biotropica 36:20–32
Larson PR (1962) A biological approach to wood quality. Tec Assoc Pulp Pap Ind 45:443–448
Larson PR (1964) Some indirect effects on wood formation. In: Zimmermann MH (ed) The formation of wood in forest trees. Academic, New York, pp 345–365
Lohidas T, Dayanand T (1984) Some distinguishing characteristics between wavy-grained and straight-grained trees of Red Sanders (*Pterocarpus santalinus* Linn.f.). Indian For 7:68–69
MacLachlan I (2009) Identifying the illegally traded Indian wood *Pterocarpus santalinus* (Leguminosae). Q J For 103(2):129–113
Mulliken T, Crofton P (2008) Review of the status, harvest, trade and management of seven Asian CITES-listed medicinal and aromatic plant species. BundesamtfürNaturschutz, Bonn. 156p
Okuma M (1998) Wood utilization in the 21st century. Wood Ind 52:98–103
Rao KR, Purkayastha SK (1972) Indian woods; their identification, properties and uses. Forest Research Institute, Dehradun. 124p
Richter HG, Dallwitz MJ (2000) Commercial timbers: descriptions, illustrations, identification, and information retrieval. http://www.delta-intkey.com/wood/en/www/papptsan.htm. Accessed 28 Sept 2018
Shukla SR, Sudheendra R, Rao RV (1999) Preliminary studies on physical and mechanical properties of plantation grown Andaman padauk (*Pterocarpus dalbergioides*) from Karnataka. J Timb Dev Assoc India 95:11–15
Simpson A, TenWolde A (1999) Wood handbook- wood as an engineering material. U.S. Department of agricultural forest service. For Prod Lab 3:5
Subasinghe S, Yoon MH (2011) *Pterocarpus santalinus* Linn. f. (Rath handun): a review of its botanical uses. Phytochem Pharmacol 24:67–72
Suresh K, Hegde M, Deenathayalan P, Kumar PK, Thangapandi M, Singh BG, Krishnakumar N (2017) Variation in heartwood formation and wood density in plantation-grown Red Sanders (*Pterocarpus santalinus*). In: Pandey K, Ramakantha V, Chauhan S, Arun Kumar A (eds) Wood is good. Springer, Singapore, pp 139–151. https://doi.org/10.1007/978-981-10-3115-1_14
Swaminathan C, Rao RV, Shashikala S (2012) Preliminary evaluation of variations in anatomical properties of *Melia dubia*. Cav. wood. Int Res J Biol Sci 1(4):1–6
Troup RS (1921) The silviculture of Indian trees, Dilleniaceae to Leguminosae (Papilionaceae). Clarendon Press, Oxford, pp 123–135
Vijendra Rao R, Chauhan L (2003) Wood anatomy of legumes of India: their identification, properties and uses. Bishen Singh Mahendra Pal Singh, Dehra Dun
Walter KS, Gillet HJ (eds) (1998) 1997 IUCN Red List of threatened plants. IUCN, Gland
White DJB, Robards AW (1996) Some effects of radial growth rate upon rays of certain ring porous hardwoods. J Inst Wood Sci 17:45–52
Wiemann MC, Williamson GB (2013) Biomass determination using wood specific gravity from increment cores. Forest Products Lab, USDA Forest Serv. 225p

Threats and Conservation 12

S. Noorunnisa Begum, K. Ravi Kumar, and B. N. Divakara

Abstract

Red sanders (*Pterocarpus santalinus*) is restricted in its distribution to Southern Eastern Ghats of India and listed by IUCN as endangered species. *P. santalinus* is the only species of *Pterocarpus* that has been listed in Appendix II of CITES. Major threats for this endangered tree are illegal and unsustainable harvesting, forest fire, grazing, loss of habitat, inbreeding, pests, diseases, low fruit set and poor regeneration. The threats and methods for conservation of this precious tree are reviewed and discussed.

Keywords

Pterocarpus santalinus · Threats · Conservation

12.1 Introduction

Pterocarpus is a renowned genus of trees and woody climbers distributed in three tropical regions, i.e. Neotropics, Tropical Africa and Indomalaya (Indian subcontinent and Malay Peninsula/Archipelago). It consists of 35–40 species (Klitgaard and Lavin 2005) but is estimated to have up to 60 species (Chauhan and Vijendra Rao 2003). *Pterocarpus* has gained popularity globally because of its unique wood features, for its ethnomedicinal uses and valuable bioactive compounds. The importance of this genus was appreciated by Saslis-Lagoudakis et al. (2011) by

S. Noorunnisa Begum (✉) · K. Ravi Kumar
Centre of Repository of Medicinal Resources, School of Conservation of Natural Resources, Foundation for Revitalization of Local Health Traditions, Bangalore, India
e-mail: noorunnisa.begum@tdu.edu.in; k.ravikumar@tdu.edu.in

B. N. Divakara
Institute of Wood Science and Technology, ICFRE, Bangalore, Karnataka, India

T. Pullaiah et al. (eds.), *Red Sanders: Silviculture and Conservation*,
https://doi.org/10.1007/978-981-13-7627-6_12

mentioning the application of community phylogenies in bioscreening, also focussed on the processes of shaping cross cultural ethnomedicinal patterns. There are only four species of *Pterocarpus* found in India, namely, *Pterocarpus dalbergioides*, *P. indicus*, *P. marsupium* and *P. santalinus*. Out of these four species, the International Union for Conservation of Nature (IUCN) has listed *P. indicus* and *P. marsupium* as vulnerable and *P. santalinus* as endangered. *P. dalbergioides*, an endemic species of Andamans, has been classified by IUCN as vulnerable (Barstow 2018b). Considering the present status Lakshminarasimhan and Mondal (2012) indicated that *P. santalinus* is critically endangered. Similarly, Prasad et al. (2008) suggested that it is appropriate to consider *P. dalbergioides* as threatened and if not protected well may swiftly move to the next category of extinction. In the natural habitats of Sri Lanka, the species of *Pterocarpus* is rare due to continuous anthropogenic pressure and harvesting the trees indiscriminately (Arunkumar 2011).

12.2 Distribution Range

Pterocarpus santalinus has a restrictive distribution in the South Eastern portion of the Indian Peninsula (Fig. 12.1 and Fig. 12.2). It is native to the states of Andhra Pradesh, Tamil Nadu and Karnataka (Hedge et al. 2012; Arunkumar and Joshi 2014). More specifically it is found in the Palakonda and Seshachalam hill ranges of Kadapa-Chittoor Districts of the State of Andhra Pradesh (Rajampet, Rayachoti, Balapalle, Kodur ranges, Gangana Palle forest of Vempalle village and Lankamala Reserve Forest) (Hedge et al. 2012; Arunkumar and Joshi 2014). It occurs less frequently in the regions of Chengalpattu and in the Vellore (formerly North Arcot)

Fig. 12.1 Forest View showing trees of Pterocarpussantalinus in Chittoor district

Fig. 12.2 India district level map showing the distribution of the *Pterocarpus santalinus*

Hills, Tamil Nadu (Hedge et al. 2012). Red sanders is endemic to the hills of Kadapa, Chittoor, Nellore, Kurnool and Prakasam districts of Eastern Ghats of Andhra Pradesh (Figs. 12.1, 12.2 and 12.3) and extending and to the adjoining regions of Tamil Nadu and Karnataka. In Tamil Nadu it was reported from the drier zones of Chengalpattu and Vellore districts. In Karnataka it has been reported from Devarayanadurga, Sandoor and Karpakapalli MPCAs. The existence of wild population in the above regions of Karnataka needs to be ascertained. It is commonly planted either as monoculture or as an ornamental tree by forest department.

The species is cultivated and in Kerala, Maharashtra, Odisha and West Bengal within India and within Sri Lanka around its wild range states.

Fig. 12.3 *Pterocarpussantalinus* in Rajampetta forest

12.2.1 Estimated Area of Occupancy (AOO)

The estimated area of occupancy (AOO) is 5160 km^2. It is found only in ten locations between 150 and 900 m elevation. The species has an estimated extent of occurrence (EOO) of around 25,000 km^2. Within the Kadapa landscape, Arunkumar and Joshi (2014) estimated the species to have a distribution of 5160 km^2. However, another estimate for Andhra Pradesh considered species distribution to not exceed 398,000 ha equivalent to 3980 km^2 (Hedge et al. 2012). The species total area of occupancy (AOO) is not anticipated to be much larger than this measure as it is most common in this landscape and in Andhra Pradesh. Currently the number of locations of the species is not fully known, but it is not predicted to be larger than ten based on the consistency of threat across the species range. The species is estimated to have an extent of occurrence (EOO) to around 25,000 km^2.

Though historically the decline of population is known, the information on the scale of this is currently unavailable and hence cannot be estimated. The population is further declining due to the illegal harvest of mature trees to cater the much sought-after timber. The species is globally assessed as near threatened as it almost qualifies for vulnerable B1ab(iii,v) category (Barstow 2018). The species is currently protected under national and international trade regulations. The focus on conservation of this species is by halting the illegal harvest and by promoting plantations and cultivated areas.

Fig. 12.4 Profile of *Pterocarpussantalinus*

12.3 Major Threats

12.3.1 Population

Pterocarpus santalinus population is under pressure and in decline due to illegal felling and human pressure on "Red Sanders" forests. In a selection of 0.1 ha sample plots, *P. santalinus* occurred at a density of 16.75 per 0.1 ha. More specifically trees above 30 cm dbh occurred at a density of 9.19/0.1 ha plots, but trees exceeding 70 cm dbh were only found at a density of 13.2 trees per ha (Rao and Raju 2002). In the same sample plots, seedlings occurred at an average density of 0.74 per 0.1 ha.

Overall, *P. santalinus* has a skewed size class distribution which consists of significantly fewer trees over 50 cm dbh compared to trees of a smaller size class. The species has been subject to historical population decline due to overextraction of trees for timber.

These declines are anticipated to be large and have been occurring over many generations of the species; however the information is not available to estimate the scale of this ancestral loss.

12.3.2 Extrinsic Factors

12.3.2.1 Habitat Specificity and Ecology

Pterocarpus santalinus is a medium-sized tree species with an average height ranging between 10 and 15 m (Arunkumar and Joshi 2014). It grows in dry deciduous forests where it can be mixed with other native species or as pure strands (Babar et al. 2012). The species has a narrow specificity, with most them growing on quartzite soil. It grows on dry, hilly zones often on rocky grand (Rao and Raju 2002). The species is light demanding and cannot tolerate shade or water logging (Arunkumar 2011). It is a slow-growing tree, taking 50–60 years to reach 70 cm girth. The trees are pollinated by bees, and fruit is wind dispersed and seed germinated quickly after the rainy season. The species occurs from 150 to 900 masl (Hedge et al. 2012).

12.3.2.2 Forest Fire

The red sanders is a very hardy species. The annual incendiary fires started by graziers, however, cause severe setback to the young seedlings.

12.3.2.3 Grazing

Young plants are readily browsed by cattle and deer, but effective damage caused by them is negligible in view of the plentiful fodder grass around (Ramakrishna 1962).

12.3.2.4 Fuel

The worst menace is man, who, in his quest for a tree containing solid heartwood preferably with wavy grain, damages other trees irretrievably, rendering them useless and fit only for fuel. The forests are surrounded by professional smugglers who seem to dictate the type of silvicultural system should be adopted (Ramakrishna 1962).

12.3.2.5 Loss of Habitat

Pterocarpus santalinus is threatened by habitat loss due to anthropogenic pressures on deciduous forests in Tamil Nadu and Andhra Pradesh.

12.3.2.6 Unsustainable Harvest

It is further aggravated by the special demand for wavy grained individuals which cannot be determined by any observable morphological parameters, causing indiscriminate felling.

12.3.2.7 Use and Trade

This species is threatened because of illegal timber extraction. Red sanders distributed near deprived villages, where the extraction and trade of red sanders wood is a main source of livelihood, is a risk to the species (Rao and Raju 2002).

Pterocarpus santalinus is a very valuable, attractive, hardwood species. Its timber is used to make furniture, musical instruments, carving and to make agricultural tools. The wood is in high demand in China and Japan but was previously sought in many parts of Europe (Arunkumar and Joshi 2014). Arunkumar and Joshi (2014) stated that "the regular seizures of huge quantities of red sanders wood in international markets are so common now that the survival and existence of *P. santalinus* is a cause for concern".

12.3.2.8 Medicinal Use

The species may also be harvested for pharmaceutical and medicinal uses. The powdered heartwood is used for treating diabetes. The tree is also used in the preparation of immune boosting medicine in China. The wood is astringent which can help alleviate swelling, pain and reduce bleeding (Rao and Raju 2002). The characteristic red dye known as "santalin" from red sanders is known for medicinal properties and is also used to dye paper, pulp, textiles and for tanning (Arunkumar 2011; Prakash et al. 2006).

12.3.2.9 Pests

Caterpillars of Pyralidae (Lepidoptera) attack red sanders poles generally at the fork of the tree (Luna 2005). Partially dead wood and dried wood are infested by Bostrychidae – *Sinoxylon anale* and *S. crissum*.

12.3.2.10 Disease

Sclerotium rolfsii causes foliage blight in *P. santalinus* (Sankaran et al. 1984).

12.3.3 Intrinsic Factors

12.3.3.1 Inbreeding

The extraction of the largest trees for timber has left the remaining population of *Pterocarpus santalinus* skewed, so there is reduced regeneration and increased occurrence of inbreeding. The illegal trade adversely affected the population structure of the species with the removal of superior phenotypes. The natural fruit set is very low, about 6%, comparative to the quantum of flowers, with xenogamous fruits alone carrying to maturity and dropping of autogamous and geitonogamies. This puts the species at future risk of further decline.

12.3.3.2 Poor Natural Regeneration

Currently, the species also suffers from low fruit set and hence poor natural regeneration. Therefore, in the future, the population of the species may decline further, and it is also predicted to suffer from genetic erosion and inbreeding depression due to a relatively small remaining population (Arunkumar 2011).

12.4 Conservation Strategies

During the recent years, the population of *Pterocarpus santalinus* is depleting at an alarming rate. In majority, species is currently observed as frequent but is treated as one of the critically endangered medicinal species of India. Causes of decrease in population of the species are directly related to illicit felling, recurrent forest fires and possibly muted regeneration in vulnerable pockets of distribution.

Thus, the species has been categorized as endangered by IUCN and has been listed in Appendix II of Convention on International Trade in Endangered Species of Wild Fauna and Flora (CITES). The CITES aims at ensuring the survival of wild animals and plants being traded internationally and thereby safeguarding the species being overexploited. *P. santalinus* is the only species of *Pterocarpus* that has been listed in Appendix II of CITES on 16th February 1995, on proposal from India. Appendix II includes all species which although not necessarily now threatened with extinction may become so unless trade in specimens of such species is subject to strict regulation to avoid utilization incompatible with their survival. The listing was annotated to cover – logs, woodchips and unprocessed broken material which subsequently modified at CITES COP 14 as logs, woodchips, powder and extracts. In Andhra Pradesh, it is protected under the red sanders and Sandalwood Transit Rules of the Andhra Pradesh Forest Act, 1967, and is also classified as a "reserved tree" under the Andhra Pradesh Preservation of Private Forest Rules, 1978.

12.4.1 Schemes to Encourage Red Sanders Cultivation

From the year 2011, the State Government launched Krishi Aranya Protsaha Yojane scheme to enlist the public and cooperation of farmers for increasing tree cover including red sanders cultivation. In the scheme farmers, public and NGOs can obtain the seedlings from nearest nurseries of the department at subsidized rates, and incentives are given to farmers to encourage them to grow native tree species on their lands. Forest department provided each seedling at incentive rate of Rs. 10 to Rs. 20 per seedling. Assessment of the seedlings would be done by the respective forest officers (http://aranya.gov.in/downloads/Planting-Techniques.pdf).

12.4.2 Awareness Campaigns

Organizing awareness campaigns with respect to protection of trees especially red sanders and its natural habitat with public cooperation can ensure an effective way to control the illegal activities. Messages such as "Red sanders is wealth of the people available in our forests" and "The rich and powerful are getting crores while poor people are killed and jailed due to the smuggling – Don't involve in the smuggling" were used during the campaigns of the Andhra Pradesh police and Forest officers.

People were also educated on Forest Rights Act (FRA), to ensure government scheme for providing assistance for livelihood and land rights, in order to ensure protection and regeneration of red sanders. Promotion was done involving people living in and around the forests and by creating livelihoods as Red Sander-based skilled workers. (http://www.academia.edu/10525504/Red_Sanders_Smuggling_causes_and_consequences).

Anuradha in the project supported by Rufford Foundation, United Kingdom (www.rufford.org/rsg/projects/anuradha_maniyam), created awareness about red sanders, importance of plantation, distribution of saplings and efforts to conserve the tree by involving in local communities.

12.4.3 Promoting Red Sanders Based Industries Within the Country

Few suggestions for supporting the red sander-based small- and large-scale industries in the country which in turn helps economic growth of the country are:

1. "A" grade red sanders can be utilized for trade at international level, especially with China – the largest consumer. Industries can be set up as joint venture in association with China. Employment can be provided for local people, skilled at management of Red Sander-based industrial processes through technology transfer on and profit sharing basis from China.
2. Utilization of "B" grade red sanders for furniture to trade within the country and export to consuming countries.
3. The carvings and art made from "C" grade red sanders have great markets in both local and international markets by distributing to local artists to create job opportunities.

Hence smuggling and other illegal activities can be curbed to the maximum extent, and employment opportunities for the local people will be enhanced (http://www.academia.edu/10525504/Red_Sanders_Smuggling_causes_and_consequences).

12.4.4 Acts to Protect/Conserve Red Sanders

The Andhra Pradesh Sandalwood and red sanders Wood Transit Rules, 1969, and Andhra Pradesh red sanders Wood Possession Rules, 1989, laid down the following conditions for import, export and movement:

1. No person shall import Sandalwood, Sandalwood chips, Sandalwood powder or Red Sanders wood, Red Sanders woodchips and Red Sanders wood powder into or export Sandalwood chips, Sandalwood powder or Red Sanders wood, Red

Sanders woodchips and Red Sanders wood powder from or move Sandalwood or Red Sanders wood within any place in the whole of the State, unless such Sandalwood or Red Sanders wood is accompanied by a permit.

2. No piece of the wood and other materials specified such as bags containing saw dust or chips of such wood shall be transported unless they bear an authorized marking or seal of the Government (Mahammadh 2014).
3. According to Andhra Pradesh Forest Offences (Compounding and Prosecution) Rules – 1969:
   (a) The forest produce seized may be kept in the safe custody of the Forest Guard or Forest Watcher, and a certificate to that effect shall be obtained.
   (b) The report of seizure of the forest produce shall be submitted immediately, to the next superior officer and Divisional Forest Officer.
   (c) An enquiry into a forest offence shall be held by an officer not below the rank of a "Deputy Range Officer or Forester".
   (d) Every accused who expresses his willingness to have the offence compounded shall forthwith give a written undertaking in that regard.
   (e) In case the value of forest produce seized is less than 2 [Rs. 50,000 (Rupees Fifty Thousand)] and where the accused does not or compounding the offence, or the competent authority does not choose to compound the offences, the Forest Range Officer may order the prosecution of the accused. In all other cases, where the accused does not choose to compound the offence, the Sub-Divisional Forest Officer, Managers of Primary Marketing Societies under the administrative control of Girijan Cooperative Corporation Limited or the Divisional Forest Officer may order the prosecution of the accused (Bhagyaraj and Ramana 2013).

12.4.5 Mandate of Trade Policies

The Convention on International Trade in Endangered Species of Wild Fauna and Flora (CITES), in combination with the Export and Import Policy (EXIM) through Foreign Trade (Development and Regulation) Act 1992 and enforced via the Customs Act (CITES Management Authority of India, 2004) and Wildlife Protection Act, 1972/1991/2002, is implemented in India.

Red Sanders are placed in Appendix II of CITES in 1995. But, it requires a scientific study of the species as Non-Detriment Finding (NDF). The rules of the CITES make it a mandate for NDF be done by the Scientific Authority (SA) of the country. The NDF reports determine the further use of resources, or it should be used in sustainable manner for commercial international trade without damaging the growth and abundance of the species in the wild. However, the continuous logging and illegal trade can led CITES to imposing an absolute ban in the international trade for red sanders in India (http://www.academia.edu/10525504/Red_Sanders_Smuggling_causes_and_consequences).

12.4.6 Import and Export Government Rules

The main laws governing harvesting of medicinal plants in India are the Indian Forest Act (1927) and, to a lesser extent, the Wildlife (Protection) Act (1927/1991/2002). The Indian Forest Act (1927) consolidates the law relating to forest produce, the transit thereof and duty thereon, and empowers State Governments to regulate the transit of forest produce, e.g. medicinal plants. The act deals specifically with reserved, protected and village forests. Almost all the States and Union territories in India have regulations regarding harvest, transit and trade in medicinal plants. Most have established lists of species banned from harvest from forests (Negative Lists), which include threatened plants (Mulliken and Crofton 2008).

The Indian Forest Act (1927) has been adopted by most of the States and is directly applicable to the Union Territories of India. The remaining States have enacted state forest acts of their own, which are largely based on the Indian forest act. The forest acts of the States have been amended from time to time. The States have framed rules under the acts to protect and preserve the forest wealth of their respective States.

Pterocarpus santalinus is classified as a "reserved tree" under the Andhra Pradesh Preservation of Private Forest Rules, 1978. Cutting, transport and sale require permission from the Divisional Forest Officer in accordance with rules set by the State Government. According to a State Forest Officer, felling of this forest species has been banned since 1982. The Andhra Pradesh Sandalwood and Red Sanders wood Transit Rules (1969) also specify that any import, export or transport of *P. santalinus* wood, chips or powder must be accompanied by a permit detailing the items and quantities involved and their source and destination. Further, the rules require that all items in trade (including individual wood pieces, bags of powder, etc.) be marked and, if relevant, sealed. There are also provisions for the marking of the individual trees at the time of felling and onward chain of custody requirements. The Andhra Pradesh Red Sanders possession rules require a license to be obtained for possession of *P. santalinus* more than a certain amount and the manufacture and/or trade of *P. santalinus* products and to keep detailed records of stocks, transactions and storage facilities. This rule does not apply to domestic use. The Government of Andhra Pradesh has considered charging *P. santalinus* smugglers under the Preventive Detention Act. The Andhra Pradesh forest department has initiated a programme to improve regeneration of *P. santalinus* and their harvestable stock in natural forest areas through fire management, weed suppression and other activities (Mulliken and Crofton 2008).

The Tamil Nadu Red Sanders wood transit rules (1968) also specify that any import, export or transport of *P. santalinus* wood, chips or power must be accompanied by a permit detailing the items and quantities involved and their source and destination. No person shall have in his possession move red sanders timber chips or powder except under a special permit issued by the district forest officer in form VI, provided that nothing contained in the rules shall apply to red sanders wood not exceeding 5 kg in weight carried by any bona fide traveller or any person authorities by him in writing for his personal use, from trees growing on the patta lands of the said farmer or of any neighbouring farmer from whom title to such Red sanders wood is derived and provided that he obtained a certificate of title from the village munsif and that the quantity cut and carried on each occasion does not exceed 5 kg in weight.

Possession and transport of *P. santalinus* timber, chips and powder similarly require permits from the District Forest Officer in the State of Tamil Nadu but only if the quantity involved exceeds five kilogrammes. Felling of this species is currently banned in Tamil Nadu. There are also restrictions and government permissions required for the cutting of trees in hill areas and provision made for dictating the right to fell trees on private lands. However, in Karnataka, no specific rules appear to be there with respect to *P. santalinus*, but the species is subject to detailed permit requirements for transport and sale, along with all other "forest products". This means that the products are to be sold via auction, tender or "tender-cum-auction", with the permission of the Panchayat. Harvest in district and protected forests requires the permission of the Forest Officers (Mulliken and Crofton 2008).

12.4.7 Suggestions for Curbing the Illegal Logging and Smuggling

The smuggling cannot be curbed unless and until the Adivasis and the forest dwellers of Tamil Nadu and Andhra Pradesh are scientifically addressed. The policies should be designed in a way to discourage poor and local people from participating in smuggling and other illegal activities of the forest and to encourage them in actively taking part in maintenance of Red Sandalwood natural regeneration and forest resources (http://www.academia.edu/10525504/Red_Sanders_Smuggling_causes_and_consequences):

- The main drivers of illegal logging and associated trade should be identified to design the control measures by providing transparency on land tenure and by establishing legal origin of timber and legal compliance in forest management unit.
- The forest information systems should be designed for improved governance like accountability and transparency of both private sector operators and forest authorities.
- The priority should be given to the land rights of indigenous people and community customer land rights, beyond the scope of forest projects.
- The suggestions should be taken from national NGOs, indigenous peoples' organizations and private sectors for effective forest governance and certification.
- Undertaking the valuable technical studies and projects can enhance the circulation of information to stakeholders, private sector, NGOs and other government agencies and universities. The dissemination of information on policies to public through newspapers and other media plays a vital role.

12.4.8 Ex Situ Conservation

Studies on vegetative propagation by Reddy and Srivasuki (1990) have summarized the various vegetative propagation methods such as grafting, air layering and rooting of cuttings carried out by the research wing of Andhra Pradesh Forest Department. Micropropagation studies on red sanders began in the 1980s. Various researchers used different explants to regenerate plantlets. Sarita et al. (1988) used single node and terminal explants for multiplying plantlets. Sita et al. (1992) induced shoots from shoot tips and successfully transferred micropropagated plants to field. Anuradha and Pullaiah (1999) reported highest shoot bud regeneration by culturing mesocotyl explants. Arockiasamy et al. (2000) found that detached cotyledons from in vitro germinated seedlings were useful, while Prakash et al. (2006) used nodal explants of young shoots. In vitro germination studies carried out by Chaturani et al. (2006) indicated that Anderson medium was ideal for its establishment and suggested that the pods must be used within 1 week after harvest to obtain higher germination percentage. Rajeshwari and Paliwal (2008) reported successful micropropagation protocol through axillary shoot proliferation by using cotyledonary nodes. Nodal segments from in vitro regenerated shoots obtained from mature trees proliferated into multiple shoots (Padmalatha and Prasad, 2008). Balaraju et al. (2011) induced multiple shoots by using shoot tip explants derived from 20 days old in vivo germinated seedlings and confirmed the genetic fidelity. Warakagoda and Subasinghe (2013) studied different explants and acclimatized the in vitro rooted plants. These optimized techniques by various researchers can be used for mass multiplication through micropropagation.

12.4.9 In Situ Conservation

Red sanders figures in Appendix II of CITES Status as well as the negative list of exports notified by Govt. of India (Notification 2 (RE-98) dt 13.4.98, 1997–2002).

The species is present in six ex situ collections (BGCI 2018). The species is considered endangered in India and is protected under law. Original laws prohibiting the harvest of the species were established in the nineteenth century, and they have been developed over the last century. The species has been internationally protected under CITES Appendix II since 1995. Within Andhra Pradesh, it is estimated that 168,000 ha of red sanders forest is found within protected areas.

There is no harvest management for *P. santalinus* from wild populations as any extraction of the species is illegal, so current management plans focus on preventing the illegal harvest and trade of the species. There is encouragement to raise plantations of *P. santalinus* outside of its natural range and in private plantations. It is also recommended to streamline the process of exporting and harvesting of propagated and cultivated trees to encourage the sustainable use of these trees. There are some in situ conservation efforts of remaining red sanders forests. To achieve this there is a need for protection from fires and cattle grazing as well as the need for planting of *P. santalinus* in areas of scarcity, within its native range (Rao and Raju 2002).

At present the harvest of this species is monitored, and illegally harvested trees are seized. Legal export requires contracts and permits from multiple government agencies. Extract of living trees from the natural forests is prohibited, and silvicultural removals, if any, are as per the prescriptions of the approved working plans. In the protected areas, removals of any kind are prohibited. The species should be used in social forestry; these could be a source for sustainable harvesting. The species survives well without irrigation and fertilizer. Also, detailed studies on the reproductive biology of this species are needed to produce and maintain the genetic diversity. The continued conservation of this species is necessary for its continued survival in the wild.

12.4.10 Current Situation of Red Sanders

Trade of this timber is now currently from illegal sources as the harvest of the species is prohibited. The trade of wood from even cultivated sources is limited. Legal trade is limited to the occasional sale of confiscated timber by the Government of Andhra Pradesh. The species has negligible utilization within the country mainly in Ayurvedic pharmacopoeia and sometimes for making small toys and musical instruments. The species has virtually no domestic demand for construction or future use. Almost all the seizures indicate the movement of logs towards the exit points, or the seizures themselves are at the exit points during attempted smuggling (MoEFCC 2018). Between 1991 and 2003, a 717.7 MT of *P. santalinus* chips were traded. Most of the trade of the species is now in chips or dye. There are many regional, national and international laws (including CITES) in place to monitor and control the trade of this timber (Rao and Raju 2002). Current extraction of timber is opportunistic, and mostly of trees exceeding 50 cm in diameter, due to the illicit nature of the harvest of this species, there is no "commensurate replacement of natural stands" (Arunakumar 2011).

The International Union for Conservation of nature conservation has now reclassified red sanders (*Pterocarpus santalinus)* as "near threatened" from the earlier "endangered".

After being classified as "endangered" in 1997 and added to the Red List, this is the first time that red sanders has been shifted to a better conserved category. Red Sanders wood fetches huge prices in the international market for its use in making luxury products, musical instruments and medicine. A tonne of red sanders costs anything between Rs. 50 lakh to Rs. 1 crore in the international market.

However, as the species is endemic to India and had been considered endangered, the foreign trade policy of India doesn't allow its export. This ban, coupled with its high demand, made illegal trade of red sanders rampant in states where it is found. In fact, illegal logging is a reason for its declining population, according to the IUCN. The species has been subjected to historical population decline due to over extraction of trees for timber. These historical declines are anticipated to be large and have been occurring over many generations. However, the information is not available to estimate the scale of this ancestral loss.

Apart from focussing on the illegal felling, the IUCN has also urged a sustainable harvesting process for the species, and the recommendations are, "There is no harvest management for *Pterocarpus santalinus* from the wild populations as any extraction of the species is illegal, so current management plans focus on preventing the illegal harvest and trade of the species. There is encouragement to raise plantations of *Pterocarpus santalinus* outside its natural range and in private plantations. It is recommended that the process of exporting and harvesting be propagated, and process of cultivating trees be streamlined to encourage the establishment and use of these trees.

While the foreign trade policy doesn't allow export or red sanders wood, it lets states export confiscated timber from illegal sources. Experts say that this policy is encouraging the illegal trade as opposed to trade from plantations.

12.4.11 Policies

The dichotomy of trading in red sanders is intriguing. Though a farmer can grow the tree, he/she would require permits to fell and transport the wood, which is difficult to obtain. Moreover, the price of this wood in the domestic market is less than half of what it is in the international market as demand is low. At the same time, the farmer cannot export it as the foreign trade policy prohibits it. Ironically, the Indian government had itself asked the quotas to export red sanders from CITES as the tree is categorized as a species that needs protection. Now the government sells seized logs from smugglers to private companies for export by granting them a license.

The Directorate General of Foreign Trade (DGFT) reads: "As per the existing export policy, wood of red sanders in any form, where raw, processed or unprocessed, is prohibited from export". Only value-added products such as dyes and musical instruments made from red sanders and procured from legal sources are permitted for exports under a restricted category requiring a proper license.

12.4.11.1 Threats

Red sanders, known for its rich hue and therapeutic properties, is high in demand across Asia, particularly in China and Japan, for use in cosmetics and medicinal products as well as for making furniture, woodcraft and musical instruments. Its popularity can be gauged from the fact that a tonne of red sanders costs anything between Rs. 50 lakhs to Rs. 1 crore in the international market.

Found across the southern Eastern Ghats, the tree is endemic to several districts in Andhra Pradesh and some parts of Tamil Nadu and Karnataka. But overexploitation promoted the Union government in the 1980s to recommend inclusion of Red Sanders in Appendix of CITES, which says "Trade must be controlled in order to avoid utilisation incompatible with their survival". The species was listed in Appendix II of CITES in 1995, and subsequently export of red sanders was prohibited in 2004.

In 2010, when the CITES was planning to suspend trade of red sanders obtained from India, the government submitted a Non-Detriment Finding (NDF) report saying it must be allowed to export from cultivated sources. So in 2012, India got an export quota on red sanders from CITES, under which the country could export 310 tonnes of red sanders obtained from "artificially propagated" sources (grown on farms) and 11,806 tonnes of wood from seized sources.

12.4.12 Bizarre Logic

The result is while Maharashtra, Karnataka, Telangana, Andhra Pradesh and Tamil Nadu have organized auctions to sell confiscated red sanders, farmers are barred from exporting their produce. The ultimate beneficiaries of auctions are illegal traders and private companies. Only seized stocks of red sanders can be exported and that too when a notification is issued by the government. So while quotas regarding seized stocks have been facilitated by DGFT for export, prohibition continues to apply for stocks planted by farmers. While there is a fixed export quota under CITES for red sanders, the Indian government has prohibited export of unprocessed red sanders and does not have a fixed quota for value-added goods made from red sanders such as extracts, dyes and musical instruments.

It is significant that the NDF report only studies the population of red sanders in the wild and did not include plantations by farmers. "We have asked the Botanical Survey of India to conduct a study on both wild and cultivated red sanders", says Tarun Prathuliya who is with the Union Ministry of Environment, Forest and Climate Change. As things stand today, farmers are suffering due to the government's two-faced attitude. "Those unable to sell the wood legally are trying to recover their investment by selling their produce to illegal traders", adds Shanmugasundaram. Clearly, the government needs to find a way out of this policy quagmire.

12.4.13 Justification

Pterocarpus santalinus, an Indian endemic tree species, has a restricted geographic range in the Eastern Ghats. Historically the population was also a subject to decline due to timber demand; however, information on the scale of this is not currently available and cannot be estimated. The population is still declining due to timber demand and the illegal harvest of mature individuals to provide the much sought-after timber, red sandalwood. There is also decline in the quality and extent of the species' native habitat due to human pressures on the landscape, for example, livestock grazing. Currently the number of locations of the species is not known, but it is not predicted to be larger than ten based on the consistency of threat across the species range. The species is currently protected under national and international trade regulations. The conservation focus is on reducing the illegal harvest of the species and establishing cultivated areas of the tree to reduce risk to wild stock.

12.5 Conclusion

Considering the wood demand, restricted distribution, slow regeneration, illegal harvest, trade and habitat destruction, red sanders has been categorized as endangered by the International Union for Conservation of Nature and has been listed in the Convention on International Trade in Endangered Species of Wild Fauna and Flora and is also classified as a "reserved tree" under the Andhra Pradesh Preservation of Private Forest Rules, 1978. To revive the past glory of this valuable species, government agencies, farmers, entrepreneurs and policymakers must join hands for its protection, sustainable utilization and conservation (Arunkumar and Joshi, 2014).

12.6 Future of Red Sanders

P. santalinus is a resilient species, and its survival amidst over-exploitation from the past few centuries indicates that it is necessary to seriously think about its revival strategies. One of the best ways of conserving red sanders is not only to raise large-scale seedling-based plantations in its natural habitats but also in faraway regions having similar growing conditions which would ensure that genetic material is safe for posterity. These plantations can also act as a source of plant material for initiating further tree improvement strategies. While growing red sanders outside the forest area, it is paramount to educate the tree growers to consider the gestation of the crop. Usually, it is a tendency among growers to compare tree growing with agricultural crops, but on a long-term basis, the yield and the monetary benefits accrued by growing such valuable trees are high. The government agencies must take a lead role to encourage the farmers and entrepreneurs to grow red sanders, as Kukrety et al. (2013) suggested that it is imminent to have a sustainable wood trade policy formulated by strongly incorporating stakeholder's perceptions. Therefore, to revive the past glory of this valuable species to mankind, government agencies, farmers, entrepreneurs and policymakers must join hands together in protecting, conserving and sustainably utilizing red sanders. Though the Andhra Pradesh Forest Department is involved in the red sanders plantation in some areas, rigorous adherent steps should be taken for the proper growth and sustainability of the species, to conserve the pride tree of Andhra Pradesh (Arunkumar and Joshi 2014).

References

Anuradha M, Pullaiah T (1999) Propagation studies of Red Sanders (*Pterocarpus santalinus* L.f.) in vitro – an endangered taxon of Andhra Pradesh, India. Taiwania 44(3):311–324

Arockiasamy S, Ignacimuthu S, Melchias G (2000) Influence of growth regulators and explant type on in vitro shoot propagation and rooting of Red Sandal wood (*Pterocarpus santalinus* L.). Indian J Exp Biol 38:1270–1273

Arunkumar AN (2011) Variability studies in *Pterocarpus santalinus* in different aged plantations of Karnataka. My For 47(4):343–353

Arunkumar AN, Joshi G (2014) *Pterocarpus santalinus* (Red Sanders) an endemic, endangered tree of India: current status, improvement and the future. J Trop For Environ 4(2):1–10
Babar S, Amarnath G, Reddy CS, Jentsch A, Sudhakar S (2012) Species distribution models: ecological explanation and prediction of an endemic and endangered plant species (*Pterocarpus santalinus* L.f.). Curr Sci 102(8):1157
Balaraju K, Agastian P, Ignacimuthu S, Park K (2011) A rapid in vitro propagation of Red Sanders (*Pterocarpus santalinus* L.) using shoot tip explants. Acta Physiol Plant 33:2501–2510
Barstow M (2018a) *Pterocarpus santalinus*. The IUCN Red List of Threatened Species 2018: e. T32104A67803072. https://doi.org/10.2305/IUCN.UK.2018-1.RLTS.T32104A67803072.en. Downloaded on 22 September 2018
Barstow M (2018b) *Pterocarpus dalbergioides*. The IUCN Red List of Threatened Species 2018: e. T33261A67802958. https://doi.org/10.2305/IUCN.UK.2018-1.RLTS.T33261A67802958.en. Downloaded on 30 April 2019
Bhagyaraj A, Ramana DV (2013) Status of Red Sanders in business. Indian J Appl Res 3 (7):405–407
Botanical Gardens Conservation International (BGCI) (2018) Growing trees on the IUCN Red list
Chaturani GDG, Subasinghe S, Jayatilleke MP (2006) In vitro establishment, germination and growth performance of Red Sandalwood (*Pterocarpus santalinus* L.). Trop Agric Res Ext 9:116–130
Chauhan L, Vijendra Rao R (2003) Wood anatomy of the legumes of India. Their identification, properties and uses. Bishen Singh Mahendra Pal Singh, Dehra Dun, India
Hedge M, Singh BG, Krishnakumar N (2012) NDF report *Pterocarpus santalinus* India. Institute of Forest Genetics and Tree Breeding, pp 1–30
Klitgaard BB, Lavin M (2005) The tribe Dalbergieae *sensu lato*. In: Lewis G, Schrire B, MacKinder B, Lock M (eds) Legumes of the world. Royal Botanic Gardens Kew, London
Kukrety S, Jose S, Alavalapati JAA (2013) Exploring stakeholders' perceptions with analytic hierarchy process: a case study of Red Sanders (*Pterocarpus santalinus* L.f.) restoration in India. Restor Ecol 21(3):372–379
Lakshminarasimhan P, Mondal MS (2012) A pictorial guide to some of the Indian plants included in the CITES and negative list of exports. Botanical Survey of India, Howrah, pp 22–24
Luna RK (2005) Plantation trees. International Book distributors, Dehradun, pp 608–614
Mahammadh RV (2014) Smuggling of Red Sandal in India: a review study. Int J Econ Bus Rev 2 (10):55–63
MOEFCC (2018) https://www.dgftcom.nic.in/exim/2000/committee/efc_export_0119.pdf
Mulliken T, Crofton P (2008) Review of the status, harvest, trade and management of seven Asian CITES-listed medicinal and aromatic plant species, BfN – Skripten 227. Export Import Data Bank, Department of Commerce, India
Padmalatha K, Prasad MNV (2008) In vitro plant regeneration of *Pterocarpus santalinus* L.f (Red Sanders): an endangered medicinal plant and important timber tree. Tree For Sci Biotechnol 2:1–6
Prakash E, Shavalikhan PS, Srinivasa Rao TJV, Meru ES (2006) Micropropagation of Red sanders (*Pterocarpus santalinus* L.f.) using nodal explants. J For Res 11(5):329–335
Prasad PRC, Reddy CS, Raza SH, Dutt CBS (2008) Population structure, age of gradations and regeneration status of *Pterocarpus dalbergioides* Roxb., an endemic species of Andaman Islands, India. Pac J Sci Technol 9(2):658–664
Rajeswari V, Paliwal K (2008) In vitro plant regeneration of Red Sanders (*Pterocarpus santalinus* L.f.) from cotyledonary nodes. Indian J Biotechnol 7:541–546
Ramakrishna A (1962) The Red sanders and its future. Indian For 88(3):202–206
Rao SP, Raju AJS (2002) Pollination ecology of the Red Sanders *Pterocarpus santalinus* (Fabaceae), an endemic and endangered tree species. Curr Sci 83(9):1144–1148
Reddy KK, Srivasuki KP (1990) Vegetative propagation of Red Sanders (*Pterocarpus santalinus* Linn.f.). Indian For 116:536–540

Sankaran KV, Maria Florence EJ, Sharma JK (1984) Two new diseases of forest tree seedlings caused by *Sclerotium rolfsii* in India. Eur J For Pathol 14(4–5):318–320

Sarita S, Bhatnagar SP, Bhojwani SS (1988) Preliminary investigations on micropropagation of leguminous timber tree *Pterocarpus santalinus*. Phytomorphology 38:41–45

Saslis-Lagoudakis CH, Klitgaard BB, Forest F, Francis L, Savolainen V, Williamsons EM, Hawkins JA (2011) The use of phylogeny to interpret cross-cultural patterns in plant use and guide medicinal plant discovery: an example from *Pterocarpus* (Leguminosae). PLoS One 6(7): e22275. https://doi.org/10.1371/journal.pone.0022275

Sita LG, Sreenatha KS, Sujata S (1992) Plantlet production from shoot tip cultures of Red Sandalwood (*Pterocarpus santalinus* L.f.). Curr Sci 62:532–535

Warakagoda PS, Subasinghe S (2013) In vitro propagation of *Pterocarpus santalinus* L.f. (Red Sandalwood) through tissue culture. J Natl Sci Found Sri Lanka, 41(1): 53–63. https://www.rufford.org/rsg/projects/anuradha_maniyam. Accessed on 25.06.2018

Genetic Diversity and Conservation of *Pterocarpus santalinus* L.f. Through Molecular Approaches

13

B. K. Indu, Sudipta Kumar Mohonty, Savithri Bhat, Mallappa Kumara Swamy, and M. Anuradha

Abstract

The research and development of molecular tools in recent years has significantly influenced the tree improvement and conservation. Molecular markers, such as restriction fragment length polymorphism, random amplified polymorphic DNAs, amplified fragment length polymorphism, inter simple sequence repeat, single strand conformation polymorphism, single sequence repeats are indispensable in identifying elite lines, studying genetic diversity and phylogenetic relationships and utilizing genetic resources for crop improvement. The advancements in these DNA-based marker technologies along with high-throughput sequencing platforms have further geared plant biotechnology and have a unique role in selection of plants with desired characters. Genetic diversity though contributes to adaptability and sometimes hampers the morphological selection of plants particularly tree species. There are certain features which are expressed only after several years of juvenile growth. *Pterocarpus santalinus*, commonly known as 'red sanders', is one such tree which has elite and nonelite types. This is an endemic and endangered medicinally and commercially valued tree species with wavy grained wood and

B. K. Indu (✉)
Rishi Foundation, Bangalore, India

S. K. Mohonty · M. Anuradha
Department of Biotechnology, Padmashree Institute of Management and Sciences, Bangalore, Karnataka, India

S. Bhat
BMS College of Engineering, Bangalore, India

M. K. Swamy
Department of Crop Science, Faculty of Agriculture, University Putra Malaysia, Serdang, Selangor, Malaysia

T. Pullaiah et al. (eds.), *Red Sanders: Silviculture and Conservation*,
https://doi.org/10.1007/978-981-13-7627-6_13

smooth grained wood impregnated with different levels of santalin. The commercial significance of this tree is attributed for its wavy grained nature with deep red coloured dye santalin. Thus, precise characterization and identification of selectable markers play a crucial role for an effective conservation of *P. santalinus* and sustainable utilization. Molecular studies and marker identification are widely used for varietal identification, documentation of genetic diversity and genetic mapping and also to know about phylogenetic relationships. This chapter presents a detailed review on the molecular studies of *P. santalinus* with an emphasis on genetic conservation and diversity.

Keywords
Red sanders · Authentication · Diversity · DNA markers · RAPD · Sequencing

13.1 Introduction

Forests are an indubitable resource of global ecosystem constituting nearly 31% of the area with a wealth of 80,000–100,000 diversified tree species (Porth and El-Kassaby 2014). India, being amongst the 12 mega biodiversity countries of the world, is an asylum for many rare, endemic, precious trees, and many are threatened due to illegal exploitation. Conservation biologist's trepidation regarding loss of tree diversity as an indiscriminate felling is leading to declining of tree population. The understanding of the genetic diversity of tree species provides multifaceted advantages, such as serving as an important resource for tree breeding and improvement programmes, giving livelihood for tribes, enabling a sustainable development, and providing an excellent resource of ethnomedicine. Under this scenario, tree plant improvements, selection, domestication and their sustainable utilization are the major areas of concern for forestry and conservation.

Conserving biodiversity is the prime concept for preserving the wild gene pool. Genetic diversity plays a critical role in ensuring the survival of plant species; till date less attention has been given towards conserving tree plant genetic resources. Plant conservation approaches are usually based on general principles that lead to standardized systems to assess the extinction threats of the plant species (Moraes et al. 2014; Peñas et al. 2016). In recent times, increased efforts have been made to improve both ex situ and in situ conservation strategies which foster vital plant species and/or populations conservation (Volis and Blecher 2010; Heywood 2014; Peñas et al. 2016). Conservation approaches often need a thorough understanding of population dynamics, information about the genetic structure and genetic diversity levels within species as well as amongst population (Perez-Collazos et al. 2008; Peñas et al. 2016).

The evolution and survival of species are mainly based on their genetic diversity which makes them to adapt to changing environments and various biotic and abiotic stress factors. Hence, knowing the molecular diversity levels of any forest genetic resources can be useful in the development of efficient conservation methods and breeding approaches, which in turn enable to manage the current scenario of extinction of species. The diversity at the genome level is usually identified using several molecular markers. These molecular markers are very specific and differentiate plants from the phenotypic variations. Moreover, they can help in developing reliable and effective tools for selection and conservation methods for several endemic tree species (Jaisankar et al. 2017). Integrated approaches such as molecular markers, bioinformatics and geographical information system (GIS) are very useful in developing some suitable methods for surveying, sampling and assessing the genetic diversities (Rao and Hoskela 2001; Jaisankar et al. 2017). The major molecular markers include random amplified polymorphic DNAs (RAPDs), restriction fragment length polymorphisms (RFLPs), amplified fragment length polymorphism (AFLP), inter-simple sequence repeat (ISSR), single-stranded conformation polymorphism (SSCP) and single sequence repeats (SSRs). They are widely used to establish species or identification of varieties, identification of genetic diversity and for gene mapping and also to know about phylogenetic relationships (Jaisankar et al. 2017).

Pterocarpus santalinus L.f. (Fabaceae) is an endemic, endangered medicinally valued and commercially important tree species. It is commonly known as 'red sanders', distributed in a few areas of the southern parts of Eastern Ghats of India. The occurrence of wavy grained wood is attributed to the elite nature of this tree species which have commercial significance (Padmalatha 2005; Padmalatha and Prasad 2007; Bal et al. 2014). Due to its multiutilitarian characters, the tree is being overexploited leading to population loss in its habitat (Raju and Nagaraju 1999; Padmalatha and Prasad 2007). Thus, there is a dire need to assess its germplasm and derive conservation strategies for preserving the species. The genetic diversity is one of the crucial elements that supports survival and adaptation of a species to the environmental changes (Sun et al. 1998; Padmalatha 2005; Padmalatha and Prasad 2007). The technique of finding the genetic diversity amongst various plant species has been revolutionized and simplified by the application of molecular markers. Their simplicity, reliability, efficiency and accuracy have led their wide usage in species identification and quality control (Karp et al. 1996, 1997; Swamy and Anuradha 2011; Bal et al. 2014). This chapter summarizes the research findings on the molecular marker studies of *P. santalinus* species which is endemic to the South Indian forests.

13.2 Need for Molecular Studies in *Pterocarpus santalinus*

There are many stupendous forest trees with superlative technical qualities which are expressed very lately, probably after a decade of their plantation. The domestication and large-scale cultivation of elite trees by using seedlings are hampered invariably due to their heterozygous nature. In lieu of this, molecular studies have always been important for determination of genetic diversity within and amongst populations, verification, characterization of genotypes and marker-assisted selection (Frankham 1996; Collard and Mackill 2008). Particularly, molecular markers have proven to be nonpareil tool with various applications such as selection of clones with utilitarian traits, genetic conservation, identification of genetic diversity hotspots, genetic finger printing, barcoding, the assembly of breeding populations, gene pool monitoring, characterization of population dynamics and gene flow (Arif et al. 2011; Swamy and Anuradha 2011). Forest tree improvement programmes are benefited from this lucid technology enabling early selection of clones with elite characteristics especially late-expressing traits with reference to wood quality. One such tree species which has always been in the news, publications and concern of ecologists, conservationists and environmental scientists is *P. santalinus*, popularly known as red sanders (www.traffic.org/publications-catalogues/traffic_pub_bulletin_29_1.pdf). In Sanskrit the plant is called as 'Raktachandan' and is aptly named because of the presence of deep blood red coloured dye santalin in its heartwood. Red sanders is endemic to certain forest tracts of Andhra Pradesh and Tamil Nadu, and the distribution is restricted to selected geographical locations of Eastern Ghats of Southern India. The tree is known to be one of the luxury timbers of the world and is in high demand in both domestic and international markets. Red sanders has a wide range of applications and is a source of valuable bioactive constituents of pharmacological importance (Arunkumar and Joshi 2014). Rich in bioactive-guided principles, the tree yields quality timber, and wood is of high demand in foreign countries, especially China and Japan. The quality of the timber is attributed to its red to purple black heartwood which is extensively strong, hard, finely textured impregnated with gorgeous dye santalin, and resistance to ants and other insects (Arunkumar and Joshi 2014). The widespread demand for this tree species lured smugglers and poachers thereby making a debut by the International Union for Conservation of Nature (IUCN) under Red List as endangered species (IUCN 2014). Moreover, vulnerability to forest fires, seedling diebacks, grazing, dormancy, hard fruit coat and less germinability lead to the endangered status of this species. In addition to this, climate change and deforestation have also been some of the major reasons for loss of natural population of red sanders. It's proven that the degree of diversity gets affected by either reduction in the size of natural population or by elimination of the local population of a species (Frankham 1996; Leimu et al. 2006). With this perspective, there is a dire need to initiate and perform multifaceted studies to conserve and preserve germplasm of red sanders. Molecular studies in red sanders are cardinal for two reasons: one is to identify the presence of genes responsible for deep red and wavy grained heartwood at nursery stage, and the other one is to study and conserve the genetic diversity.

Molecular biology offers an array of technologies which can come to the rescue of plant scientists for protecting this precious tree species. All molecular markers have a unique place, and they have been proven to be instrumental tools for assessing plant's genetic resources by providing impetus to understand the distribution patterns, the extent of genetic variation within and amongst species and selection of elite varieties. Nevertheless it is very important to identify plus trees at the earlier stages, as there are certain genes which express in the later stages of life cycle determining the wood quality. In red sanders this being an important concern, many researchers investigated the possibilities of developing marker-assisted technology. Recently developed marker technologies help to understand the genetic variation in the genome in an unprecedented way. Molecular markers have diverse applications in solving many problems, but certain marker types have their limitation based on the type of genome diversity. Hence, a combination of different markers is generally acclaimed to provide a precise evaluation of the extent of inter- and intrapopulation genetic diversity of plant species that are distributed naturally. This will deliberate proper conservation directives for species that are at the risk of decline (Porth and El-Kassaby 2014).

Red sanders is no exception for high amount of genetic variations which are manifested in terms of wood quality. As reiterated from the beginning of this chapter, wood quality is determined by its wavy grain nature impregnated with scarlet red coloured dye santalin. These superior qualities of the wood occur in one in every hundred plants which is personified after 20–30 years of growth and is a challenge for commercial exploitation as this fetches high price when compared to smooth grain and lesser dye. This diversity in wood qualities may be invariably due to the genetic variations within the population and to some extent response to challenging environment. However, in red sanders this can be more of gene driven, because of the occurrence of coalesce of elite and nonelite varieties in trees of the same age growing under analogous environmental conditions, which is evinced by Indu and Anuradha (2018, unpublished data) during their survey and sample collection for the marker-assisted selection studies supported by the Department of Science and Technology, Government of India (http://online-wosa.gov.in:8080/wosa/wos/projectReceived_viewProjectRecieved.action) (2014).

The concept of molecular markers is based on the naturally occurring DNA polymorphism. An ideal molecular marker is highly polymorphic and codominant in nature. They are short DNA sequences which exhibit selective neutral behaviour and frequently occur in the genome of a species (Kundan et al. 2014). The use of DNA-based molecular markers is very advantageous in defining allele frequencies and genetic characterization of an individual species. DNA-based markers were first developed for human beings, later used in plants, and successively applied for analysing the medicinal plant genome. These molecular tools allow medicinal plant breeders to identify and select only those plants having desirable features and, thus, considerably minimize the time required for the selection. The source of majority of molecular markers are genomic DNA libraries which have been developed by some of the below mentioned methods.

Molecular markers can be categorized into different groups based on the following factors: (1) mechanisms of gene action, i.e. dominant markers; (2) the detection method, i.e. hybridization or polymerase chain reaction (PCR)-based markers; and (3) the mode of transmission, i.e. maternal or paternal organelle inheritance and maternal or biparental nuclear inheritance (Nadeem et al. 2018). Also, the advantages and disadvantages of these molecular markers are mentioned in Table 13.1.

Table 13.1 Comparison of most commonly used molecular markers

Molecular technique	Advantages	Limitations	References
Random amplified polymorphic DNA (RAPD)	Rapid, simple and inexpensive. Requires minimal amount of DNA and no sequence information	Dominant, do not distinguish between homo- and heterozygote, no potential for candidate gene mapping, low to average reproducibility	Williams et al. (1990), Grover and Sharma (2016) and Nadeem et al. (2018)
Restriction fragment length polymorphism (RFLP)	Codominant, QTL and comparative mapping potential is good, very high reproducibility	Difficult, limited by the restriction site, requires high amount of DNA	Mishra et al. (2014) and Nadeem et al. (2018)
Amplified fragment length polymorphism (AFLP)	Very good genome and QTL mapping potential, anonymous origin, medium to high reproducibility	Moderate to difficult, dominant, limited by the restriction site, average to high amount of DNA required, very limited comparative mapping potential	Vos et al. (1995), Lynch and Walsh (1998), Grover and Sharma (2016) and Nadeem et al. (2018)
Arbitrary primed PCR (AP-PCR)	Similar to RAPD except the size of the primer variation (5–8 oligonucleotides) less amount of DNA, no sequence information required, anonymous origin	Dominant, do not distinguish homo- and heterozygote, no potential for candidate gene mapping, low to average reproducibility	Welsch and McClelland (1990) and Nadeem et al. (2018)
Simple sequence repeats (SSR)	Codominant, genome and QTL mapping potential is good, medium to high reproducibility, anonymous origin	Limited by the size of the genome and number of simple sequence repeats, average to high amount of DNA required, limited comparative mapping potential	Hearne et al. (1992), Grover and Sharma (2016), Vieira et al. (2016) and Nadeem et al. (2018)
Sequence characterized amplified regions (SCARs)	Specific locus representing a single RAPD fragment, codominant, highly reproducible	Prior sequence information is required, specific primers are required	Williams et al. (1991), Grover and Sharma (2016) and Nadeem et al. (2018)
Single primer amplification reaction (SPARs)	Core motifs of microsatellite DNA	No detailed analysis can be done, similar to RAPDs	Williams et al. (1991)
Single-stranded conformational	Similar-sized DNA fragments can be	Polymorphism levels will be restricted	Orita et al. (1989) and Gasser et al. (2006)

(continued)

Table 13.1 (continued)

Molecular technique	Advantages	Limitations	References
polymorphism (SSCP)	distinguished, based on the mobility of the single-stranded DNA, cost-effective		
Sequence-tagged microsatellites (STMS)	Single locus, multiallelic, codominant, highly reproducible flanking regions of the microsatellites are used for designing primers	Prior sequence information is required, robustness is involved	Beckmann and Soller (1990)
Sequence-tagged site/expression sequence-tagged site (STS/EST)	Codominant, high reproducibility, less DNA required, good genome and QLT mapping potential and excellent candidate gene mapping potential	Limited by the number of enzyme genes and histochemical enzyme assays	Fukuoka et al. (1994)
Cleaved amplified polymorphic sequences (CAPS)	Specific for a particular locus and codominant	Limited by the restriction site, prior sequence information is required	Lyamichev et al. (1993)
Microsatellite primed polymerase chain reaction (AMR-PCR)	Primer length is varied (10–20 oligonucleotides) similar to SSRs	Limitations similar to SSR markers	Meyer et al. (1993)
Inter-simple sequence repeats (IISR)	MP-PCR with 3′ anchored primers are referred as ISSR	Polymorphism levels may be low due to absence of hyper variable regions	Hearne et al. (1992), Ng and Tan (2015) and Nadeem et al. (2018)
Amplicon length polymorphism (ALP)	More amount of variation can be detected	No specificity is observed	Ghareyazie et al. (1995) and Nadeem et al. (2018)
Allele-specific PCR (AS-PCR)	Specificity is more and is applicable in applied aspects	Wide variations cannot be detected, and it is very specific	Sarkar et al. (1990)
DNA amplified fingerprints (DAF)	Short primers (5 bp), large number of variations can be detected	No specificity observed, binds to the DNA at random	Caetano-Anolles et al. (1991)
Randomly amplified microsatellites (RAMS)	Amplification is at random, large variations which may be useful are detected	No specificity observed, binds to the DNA at random	Ender et al. (1996) and Nadeem et al. (2018)

(continued)

Table 13.1 (continued)

Molecular technique	Advantages	Limitations	References
Retrotransposon-microsatellite amplified polymorphism (REMAP)	Useful in detecting variations in the transposable elements and can be further useful in expression studies	The variation in the transposon sometimes may not be useful which may be futile	Kalender et al. (1999) and Nadeem et al. (2018)
Specific amplicon polymorphism (SAP)	Very specific and helpful in detecting the variations at specific level	Variations at random cannot be detected	Williams et al. (1991)
Single nucleotide polymorphism (SNP)	Very specific and can detect polymorphism at the nucleotide level which may be helpful in applied aspects	Not applicable for diversity analysis	Nikiforov et al. (1994), Ganal et al. (2009), Xu (2010) and Nadeem et al. (2018)
Microsatellite simple sequence length polymorphism (SSLP)	Detects variations at random with short sequences of primers	No specificity is observed, variations detected are at random	Rongwen et al. (1995)
Minisatellite simple sequence length polymorphism (SSLP)	There is less specificity as the length of primer is more	No specificity is observed, the variations are detected at random	Jarwan and Wells (1989), Ganal et al. (2009) and Grover and Sharma (2016)

13.3 Molecular Marker Studies in *Pterocarpus santalinus*

There are limited publications in molecular studies of red sanders (Table 13.2). The most important factor may be the limitation of availability of plant material, which is restricted to certain geographical locations. Apart from this, medicinal trees per se pose difficulties in molecular studies even starting from DNA isolation. For initiating any molecular marker studies, it is required to establish a reliable and easy method to isolate and purify genomic DNA with good yield. In this regard, several efforts have been carried to isolate pure DNA from *Pterocarpus* species. However, most of the reports mention the difficulty in isolating genomic DNA from *P. santalinus* as it is a tree species and contain high levels of secondary metabolites that interfere in the isolation process. The protocol optimization for DNA isolation to use in polymorphism detection technique has been reported by few researchers (Padmalatha 2005; Padmalatha and Prasad 2007; Vipranarayana et al. 2013; Lakshmi and Sriramamurthy 2013; Chaitanya et al. 2014). The yield and quality of DNA used in PCR play a crucial role in amplification. A modified CTAB method has been applied to isolate DNA from both fresh and dry leaves of *P. santalinus*. Indu and

Table 13.2 List of molecular studies performed in *Pterocarpus santalinus*

Area of study	Author (year)
Morphological and molecular diversity in *Pterocarpus santalinus* L.f.	Padmalatha (2005) and Padmalatha and Prasad (2007)
Free radical scavenging activity and HPTLC finger print of *P. santalinus* L., an in vitro study	Arokiyaraj et al. (2008)
Molecular phylogeny of *P. marsupium* and *P. santalinus*	Lakshmi and Sriramamurthy (2013)
Molecular marker study for identifying genetic relationship between quality and non-quality wood	Usha et al. (2013)
Genomic DNA isolation and purification of two endemic medicinal plants (*Pterocarpus santalinus* L. and *Pimpinella tirupatiensis*)	Vipranarayana et al. (2013)
Development of RAPD and specific scar markers for the identification of *Pterocarpus santalinus* L.	Jhansi Rani and Usha (2013)
Isolation of *Pterocarpus santalinus* L. genomic DNA, for quality check and quantification	Chaitanya et al. (2014)
Analysis of genetic diversity in the populations of *P. santalinus* detected	Chaitanya et al. (2015)

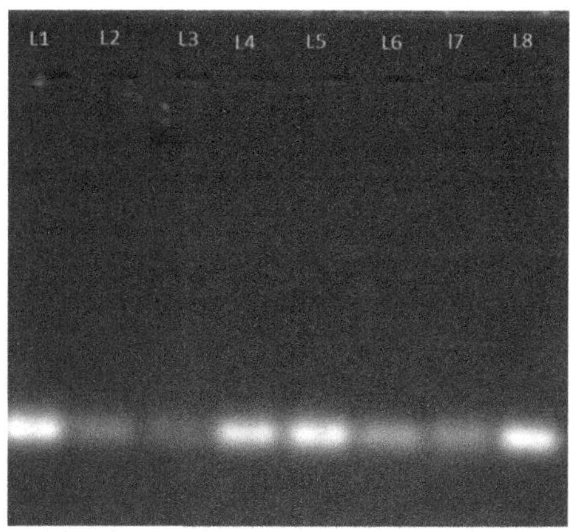

Fig. 13.1 Gel photograph of genomic DNA isolated from different samples of *Pterocarpus santalinus*. (Photo by authors)

Anuradha (2018, unpublished data) in their study identified that the higher polysaccharides, polyphenols and other secondary metabolites in freshly collected leaf sample are one of the major concerns that interferes with quality of extracted DNA. Therefore, the plantlets and leaf tissues were kept in dark at room temperature at least 12 h prior to DNA isolation to reduce polysaccharide content. The above step helped in obtaining good quality of DNA which might be due to depletion of sugar reserves and inhibition of secondary metabolite synthesis. The genomic DNA isolated from different samples was subjected to gel electrophoresis to check the quality of DNA (Fig. 13.1).

Amongst the molecular tools, RAPD has been successfully used in determining genetic diversity in various plants including the genus *Pterocarpus*. The RAPD technique was employed for the molecular characterization and identification of two species of *Pterocarpus*, i.e. *P. santalinus* and *P. marsupium*. Lakshmi and Sriramamurthy (2013) used 22 primers to screen the genome. Amongst them, ten primers gave a species-specific reproducible unique band that differentiated *Pterocarpus* species. They also studied the genetic variability between *P. santalinus* and *P. marsupium* using chloroplast marker *rbcL* gene and established the phylogenetic relationship between different species of *Pterocarpus*. Later, Usha et al. (2013) developed the RAPD profile of *P. santalinus* samples collected from the nursery and forest regions. Seven primers resulted reproducible and scorable band with high level of polymorphism with band size 1000 bp to 1.5 kb. Amongst all the primers tested, OPA8 showed the highest polymorphism in the samples evaluated. They also studied the phylogenetic relationship and observed close relationship between the forest grown and nursery grown *P. santalinus*. Similarly, RAPD markers were employed by Chaitanya et al. (2015) to evaluate *P. santalinus* genetic diversity in the region of Andhra Pradesh and Telangana states, India. They used 20 informative primers to analyse the plant diversity. The obtained results recorded the occurrence of polymorphism as revealed from the differing scorable bands. The use of unweighted pair group method with arithmetic average (UPGMA) study formed two major clusters. Also, individual plants from a population showed clustering with 75% genetic similarity. The genetic variation was found to be the highest within a population, and the lowest was found amongst populations. There was a maximum similarity percentage observed between Hyderabad and Nalgonda genotypes. Maximum similarity of 89%, minimum similarity of 61% was observed between Warangal and Chittoor region plants. A study by Padmalatha and Prasad (2007) on the morphological and molecular diversity evaluation of *P. santalinus* revealed that maximum numbers of polymorphic bands were obtained with OPC17 and OPC7 primers. Despite the commercial values, the correct identification of superior quality *P. santalinus* at their seedling stage is challenging. As mentioned earlier, the good quality aspects of *P. santalinus* can be detectable only after ~25 years of its growth and, thus, pose a major challenge for its selection or improvement through breeding. Alternatively, the use of molecular markers can be very useful to identify elite plants at their early stages of development. In this regard, Jhansi Rani and Usha (2013) developed sequence characterized amplified region (SCAR) marker to commercially identify quality plants at their seedling stages. Their study found that 2 primers (OPA-4 and OPA-20) out of 80 RAPD primers screened gave unique bands for quality plants, and hence, they were used for developing SCAR markers. They created SCAR marker by cloning and sequencing these specific unique bands. These primers were aimed to multiply unique bands (1400 and 1000 bp) and designated as SCAR primers (PTS 10 and RES 14). The polymerase chain reaction performed with SCAR marker, PTS 10, for mixed population at seedling stages significantly produced a single and sharp band (1 kb) in all quality plants. However, in non-quality plants, there was no amplification observed. Therefore, PTS 10 marker was recommended for identifying quality *P. santalinus* plants at their early growth stages.

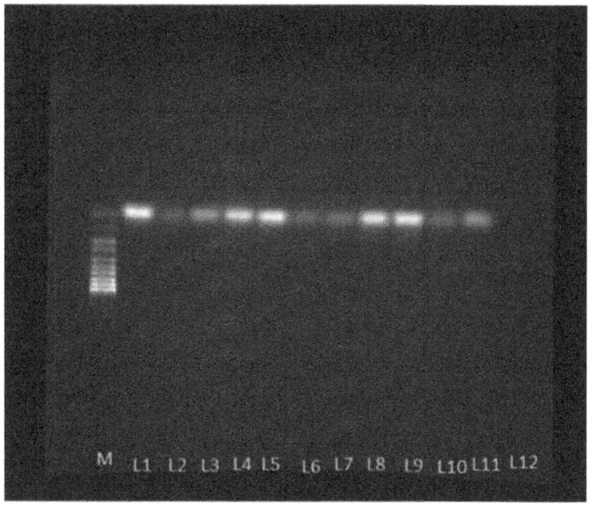

Fig. 13.2 Amplification of *rbcL* gene and *matK* genome region of *Pterocarpus santalinus* L1-L12, 100 bp molecular marker M. (Photo by authors)

In a study supported by the Department of Science and Technology, Government of India, Indu and Anuradha (2018, unpublished data) attempted to identify unique marker through DNA barcoding. DNA barcoding is a novel method for identification of plant species which also reveals genetic diversity, discrimination between elite and nonelite cultivars and phylogenetic relationship. However standardizing specific barcoding system for a plant species remains a challenging task due to high level of diversity in plant genome. In this study they have targeted two leading plastid DNA regions (*rbcL* gene and *matK* gene) either individually or in combination of these two loci (*rbcL+matK*) to enable more accurate and high-level discrimination between cultivars (Fig. 13.2.) Specific sequence of ITS (*P. santalinus* SBB-1193 and SBB-1424) was obtained through NCBI and primers designed Primer-BLAST. The result showed either presence or absence of single distinct band in two isolates (Unpublished data, Indu and Anuradha 2018).

Jiao et al. (2018) explored the use of DNA barcode for authentication of six commercially important *Pterocarpus* plant at species level. They also determined the feasibility of building wood DNA barcode reference library using herbarium specimens. Four DNA barcodes (ITS2, *matK*, *ndhF-rpl32*, and *rbcL*) either individual or in combination were investigated for their ability to discriminate amongst the species of *Pterocarpus* species. They concluded that combination of *matK+ndhF-rpl32*+ITS2 barcode shows the best discrimination potential amongst six *Pterocarpus* species, and *ndhF-rpl32* barcode performed well in distinguishing *P. santalinus*.

13.4 Conclusion and Prospects

RAPD marker is the most widely explored molecular marker for the molecular characterization and identification of two species of *Pterocarpus*, i.e. *P. santalinus* and *P. marsupium*. The genetic variability between *P. santalinus* and *P. marsupium* was also established using chloroplast marker *rbcL* gene, which also found helpful in assessing the phylogenetic relationship between different species of *Pterocarpus*. SCAR marker has been developed to commercially identify the quality of *P. santalinus* plants at their seedling stages. Further, DNA barcodes serve to authenticate *Pterocarpus* plant species and to check the quality of plantlets. Despite notable progress made in *Pterocarpus* tissue culture and morphological and biochemical studies, much of the work remains to be done to identify suitable markers for categorizing elite trees. Discovering and annotating all the genes of the *Pterocarpus* genome would lead to comprehensive understanding of the species. This could be achieved by determining the complete DNA sequence of the entire genome. The construction of genome, quantitative trait locus (QTL) and comparative and consensus linkage maps for *Pterocarpus* species will be extremely useful tools for identifying genes controlling the interesting phenotypes. Thus, appropriate marker identification in *Pterocarpus* will not only help in identifying and maintaining genetic diversity but also in conserving the germplasm of this endangered plant. Forest genomics has a bright future and awaits exiting applications in forest tree management and gene conservation. These DNA markers are unquestionably treasured tools to address plant breeding issues and population genetics. However, to reconstruct the phylogeny and taxonomic identification, they might be unpredictable and occasionally mislead, so they should be utilized with care. Future research should be encouraged towards exploring other unused DNA markers and next-generation sequencing platforms in *P. santalinus*. This will be helpful in the authentication, germplasm conservation and genetic diversity evaluation and assist breeding programmes in *P. santalinus*. These molecular markers may serve as fingerprints, which can solve the problem of *P. santalinus* adulteration with the spurious materials and impart high value to the herbal medicine.

Acknowledgements The first author Indu BK thanks the Department of Science and Technology, Government of India, for the financial support under DST (WOS-A) scheme.

References

Arif IA, Khan HA, Bahkali AH, Homaidan AA, Farhan AH, Al Sadoon M, Shobrak M (2011) DNA marker technology for wildlife conservation. Saudi J Biol Sci 18(3):219–225. https://doi.org/10.1016/j.sjbs.2011.03.002

Arokiyaraj S, Martin S, Perinbam K, Marie Arockianathan P, Beatrice V (2008) Free radical scavenging activity and HPTLC finger print of *Pterocarpus santalinus* L. – an *in vitro* study. Indian J Sci Technol 1(7):1–3

Arunkumar AN, Joshi G (2014) *Pterocarpus santalinus* (Red Sanders) an endemic, endangered tree of India: current status, improvement and the future. J Trop For Environ 4:1–10

Bal P, Panda PC, Mohapatra UB (2014) Genetic diversity and phylogenetic relationships in *Pterocarpus* species and its closely related genus *Tipuana* (Fabaceae) as revealed by RAPD and ISSR markers. Plant Sci Res 36(1&2):68–76

Beckmann J, Soller M (1990) Towards a unified approach to genetic mapping of eukaryotes based on sequence tagged microsatellites. Biotechnology 8:930–932

Caetano-Anolles G, Bassam BJ, Gresshoff PM (1991) DNA amplification fingerprinting using very short arbitrary oligonucleotide primers. Biotechnology 9:553–557

Chaitanya PJ, Chandrashekar R, Bhavani NL (2014) Isolation of *Pterocarpus santalinus* L genomic DNA, for quality check and quantification with reference to Telangana region, Andhra Pradesh, India. Indian J Sci 8(19):21–24

Chaitanya PJ, Chandrashekar R, Bhavani NL (2015) Analysis of genetic diversity in the populations of *Pterocarpus santalinus* L. detected by Random Amplified Polymorphic (RAPD) DNA markers. J Pharm Biol 5(1):1–6

Collard BC, Mackill DJ (2008) Marker-assisted selection: an approach for precision plant breeding in the twenty-first century. Phil Trans R Soc B Biol Sci 363(1491):557–572

Ender A, Schwenk K, Stadler T, Streit B, Schierwater B (1996) Rapid identification of microsatellites in *Daphnia*. Mol Ecol 5:437–447

Frankham R (1996) Relationship of genetic variation to population size in wildlife. Conserv Biol 10:1500–1508

Fukuoka S, Inoue T, Miyao A, Monna L, Zhong HS, Sasaki T, Minobe Y (1994) Mapping of sequence tagged sites in rice by single strand confirmation polymorphism. DNA Res 1:271–275

Ganal MW, Altmann T, Röder MS (2009) SNP identification in crop plants. Curr Opin Plant Biol 12(2):211–217

Gasser RB, Hu M, Chilton NB, Campbell BE, Jex AJ, Otranto D, Cafarchia C, Beveridge I, Zhu X (2006) Single-strand conformation polymorphism (SSCP) for the analysis of genetic variation. Nat Protoc 1(6):3121–3128

Ghareyazie B, Huang N, Second G, Bennet J, Kuhush GS (1995) Classification of rice germplasm I. Analysis using ALP and PCR-based RFLP. Theor Appl Genet 91:218–227

Grover A, Sharma PC (2016) Development and use of molecular markers: past and present. Crit Rev Biotechnol 36(2):290–302

Hearne CM, Ghosh S, Todd JA (1992) Microsatellites for linkage analysis of genetic traits. Trends Genet 8:288–294

Heywood VH (2014) An overview of in situ conservation of plant species in the Mediterranean. Flora Mediterr 24:5–24. https://doi.org/10.7320/FlMedit24.005

Indu BK, Anuradha M (2018) Optimization of marker assisted technology for screening, selection and micropropagation of elite genotypes of *Pterocarpus santalinus* (Red Sandalwood). Project report submitted to Dept. Sci. Tech., Government of India (unpublished). http://online-wosa.gov.in:8080/wosa/wos/projectReceived_viewProjectRecieved.action (2014) 75

IUCN (2014) IUCN Red list of threatened plants. Species survival commission IUCN. Gland, Switzerland and Cambridge, UK

Jaisankar I, Subramani T, Velmurugan A, Singh AK (2017) Genetic diversity revealed among Rattan genotypes from Andaman and Nicobar based on RAPD and ISSR markers. For Res Eng Int J 1(2):45–50

Jarwan AP, Wells RA (1989) Hypervariable minisatellites: recombinators or innocent bystanders? Trends Genet 5:367–371

Jhansi Rani S, Usha R (2013) Development of RAPD and specific scar markers for the identification of *Pterocarpus santalinus* L. J Cell Tissue Res 13(3):3809–3816

Jiao L, Yu M, Wiedenhoeft AC, He T, Li J et al (2018) DNA barcode authentication and library development for the wood of six commercial *Pterocarpus* species: the critical role of Xylarium specimens. Sci Rep 8:1945. https://doi.org/10.1038/s41598-018-20381-6

Kalender R, Grob T, Regina M, Suoniemi A, Schulman A (1999) IRAP and REMAP: two new retrotransposon-based DNA fingerprinting techniques. Theor Appl Genet 98:704–711

Karp A, Seberg O, Buiatti M (1996) Molecular techniques in the assessment of botanical diversity. Ann Bot 78:143–149

Karp AKS, Bhat KV, Ayad WG, Hodgkin T (1997) Molecular tools in plant genetic resources conservation: a guide to the technologies. International Plant Genetic Resources Institute, Rome. IPGRI Technical Bulletin No. 2

Kundan KM, Fougat RS, Ballani AT, Thakur V, Jha Y, Bora M (2014) Potential and application of molecular markers techniques for plant genome analysis. Int J Pure Appl Biosci 2(1):169–188

Lakshmi TBV, Sriramamurthy K (2013) Molecular phylogeny of *Pterocarpus marsupium* and *Pterocarpus santalinus*. World J Pharm Pharm Sci 2(6):6668–6681

Leimu R, Mutikainen P, Koricheva J, Fischer M (2006) How general are positive relationships between plant population size, fitness and genetic variation? J Ecol 94:942–952

Lyamichev V, Brow MA, Dahlberg JE (1993) Structure specific endonucleotide cleavage of nucleic acids by eubacterial DNA polymerases. Science 260:778–783

Lynch M, Walsh B (1998) Genetics and analysis of quantitative traits. Sinauer, Sunderland

Meyer W, Mitchell TG, Freedmann EZ, Vilgalys R (1993) Hybridisation probes for conventional DNA fingerprinting used as a single primers in the polymerase chain reaction to distinguish strains of *Cryptococcus neoformans*. J Clin Biol 31:2274–2280

Mishra KK, Fougat RS, Ballani A, Thakur V, Jha Y, Madhumati B (2014) Potential and application of molecular markers techniques for plant genome analysis. Int J Pure Appl Biosci 2(1):169–188

Moraes VA, Haddad D, Craessaerts K, De Bock PJ, Swerts J, Vilain S, Aerts L, Overbergh L, Grünewald A, Seibler P, Klein C, Gevaert K, Verstreken P, De Strooper B (2014) PINK1 loss-of-function mutations affect mitochondrial complex I activity via NdufA10 ubiquinone uncoupling. Science 344(6180):203–207

Nadeem MA, Nawaz MA, Shahid MQ, Dgan Y, Comertpay G, Yildiz M, Hatipoglu R, Ahmad F, Alsaleh A, Labhne N, Ozkan H (2018) DNA molecular markers in plant breeding. Current status and recent advancements in genomic selection and genome editing. Biotechnol Biotechnol Equip 32(2):261–285

Ng WL, Tan SG (2015) Inter-simple sequence repeat (ISSR) markers: are we doing it right? ASM Sci J 9:30–39

Nikiforov TT, Rendle RB, Goelet P, Rogers YH, Kotewicz ML, Anderson S, Trainor GL, Knapp MR (1994) Genetic bit analysis a solid phase method for typing single nucleotide polymorphisms. Nucleic Acids Res 22:4167–4175

Orita M, Suzuki Y, Sekiya T, Hayashi K (1989) Rapid and sensitive detection of point mutations and DNA polymorphism using the polymerase chain reaction. Genomics 5:874–879

Padmalatha K (2005) Ex-situ conservation molecular diversity of *Pterocarpus santalinus* L.f. and *Rauvolfia serpentina* (L.) Benth. ex Kurz. using RAPD markers: endangered medicinal plants. Ph.D. thesis, University of Hyderabad, Hyderabad

Padmalatha K, Prasad MNV (2007) Morphological and molecular diversity in *Pterocarpus santalinus* L.f – an endemic and endangered medicinal plant. Med Aromat Plant Sci Biotechnol 1(2):263–273

Peñas J, Barrios S, Bobo-Pinilla J, Lorite J, Martínez-Ortega MM (2016) Designing conservation strategies to preserve the genetic diversity of *Astragalus edulis* Bunge, an endangered species from western Mediterranean region. Peer J 4:e1474. https://doi.org/10.7717/peerj.1474

Perez-Collazos E, Segarra-Moragues JG, Catalan P (2008) Two approaches for the selection of relevant genetic units for conservation in the narrow European endemic steppe plant *Boleum asperum* (Brassicaceae). Biol J Linn Soc 94:341–354

Porth I, El-Kassaby YA (2014) Assessment of the genetic diversity in forest tree populations using molecular markers. Diversity. 4 6(2):283–295

Raju KK, Nagaraju A (1999) Geobotany of Red Sanders (*Pterocarpus santalinus*) – a case study from the south-eastern portion of Andhra Pradesh. Environ Geol 37(4):340–344

Rao R, Hoskela J (2001) Actions plan and research need to conserve forest genetic resources in Asia. In: Uma Shaanker R, Ganeshaiah KN et al (eds) Forest genetic resources: status, threats and conservation strategies. Oxford & IBH Publishing Company Private Limited, Calcutta

Rongwen J, Akkaya MS, Bhagwat AA, Lavi U, Cregan PV (1995) The use of microsatellite DNA markers for soybean genotype identification. Theor Appl Genet 90:43–48

Sarkar G, Cassady J, Bottema CDK, Sommer SS (1990) Characterization of polymerase chain reaction amplification of specific alleles. Anal Biochem 186:64–68

Sun M, Wong KC, Lee JSY (1998) Reproduction biology and population genetic structure of *Kandelia candel* (Rhizophoraceae) a viviparous mangrove species. Am J Bot 85:1631–1637

Swamy MK, Anuradha M (2011) Analysis of genetic variability in Patchouli cultivars (*Pogostemon cablin* Benth.) by using RAPD markers. Res Biotechnol 2(6):64–71

Usha R, Jhansi Rani S, Geethaprasuna T (2013) Genetic relationship between quality and non quality wood of *Pterocarpus santalinus* (Red Sanders) an endemic tree species by using molecular markers. J Chem Pharm Sci 6(3):189–194

Vieira ML, Santini L, Diniz AL, Munhoz CD (2016) Microsatellite markers: what they mean and why they are so useful. Genet Mol Biol 39(3):312–328

Vipranarayana S, Prasad TNVKV, Rajinikanth A, Damodharam T (2013) Genomic DNA isolation and purification of two endemic medicinal plants (*Pterocarpus santalinus* Linn.f. & *Pimpinella tirupatiensis* Bal. & Subr.) of Seshachalam Hills, Tirumala. J Biol Life Sci 4(1):415–420

Volis S, Blecher M (2010) Quasi in situ: a bridge between ex situ and in situ conservation of plants. Biodivers Conserv 19:2441–2454. https://doi.org/10.1007/s10531-010-9849-2

Vos P, Hogers R, Bleeker M, Reijans M, van de Lee T, Hornes M, Frijters A, Pot J, Peleman J, Kuiper M et al (1995) AFLP: a new technique for DNA fingerprinting. Nucleic Acids Res 23 (21):4407–4414

Welsh J, McClelland M (1990) Fingerprinting genomes using PCR with arbitrary primers. Nucleic Acids Res 18(24):7213–7218

Williams JGK, Kubelik AR, Livak KJ, Rafalski JA, Tingey SV (1990) DNA polymorphism amplified by arbitrary primers are useful as genetic markers. Nucleic Acid Res 18:6531–6535

Williams MNV, Pande N, Nair S, Mohan M, Bennet J (1991) Restriction fragment length polymorphism analysis of polymerase chain reaction products amplified from mapped loci of rice (*Oryza sativa* L.) Genome DNA. Theor Appl Genet 82:489–498

Xu Y (2010) Molecular plant breeding. CABI, Wallingford

Trade, Commerce and Socio-economic Status of Red Sanders

14

S. Balasubramanya, B. K. Indu, and M. Anuradha

Abstract

Pterocarpus santalinus, popularly known as red sanders, is much talked about species, in view of its economic importance, status, trade and smuggling activities which rammed the tree into the vulnerable status. Due to its endemic nature, slow growth, restricted commercial cultivation activities and high demand promoting to illegal felling, the population dynamics is imbalanced. To protect, conserve and promote sustainable utilization, there is a dire need for proper understanding of trade, commerce, socio-economic status and restoration efforts, and these are discussed in this chapter.

Keywords

Red sanders · *Pterocarpus santalinus* · Trade · Commerce · Smuggling · Restoration · Vulnerability

S. Balasubramanya (✉)
Department of IT, BT and S&T, Government of Karnataka, Karnataka Innovation & Technology Society, Bangalore, India

B. K. Indu
Rishi Foundation, Bangalore, India

M. Anuradha
Department of Biotechnology, Padmashree Institute of Management and Sciences, Bangalore, Karnataka, India

T. Pullaiah et al. (eds.), *Red Sanders: Silviculture and Conservation*,
https://doi.org/10.1007/978-981-13-7627-6_14

14.1 Introduction

India, one of the opulent biodiversity countries in the globe, is espoused with many fragile and sensitive ecosystems witnessing loss of innumerable species. Its diverse forest ecosystems spread over 69 million hectares of forest and tree cover (Kukrety 2011) with enthralling flora and fauna. Most of the Indian population either directly or indirectly depend on agriculture and related activities for their livelihood. People of India exhibit poverty-driven dependence on natural resources, explicitly for fodder and firewood, leading to its devastation of its forest cover. Anthropogenic pressures including unrestricted grazing, forest fires and unsustainable practices added to this dire situation. India is an abode of numerous endemic species, and one of the most discussed tree is red sanders botanically named as *Pterocarpus santalinus*. Red sanders, native of certain forest tracts of Andhra Pradesh, is one of the rich biodiversity states of the country. *P. santalinus* yields one of the finest luxury woods internationally and is Red listed due its over-exploitation resulting in reduction in population. This tree is an irresistible temptation to the smuggler, and criminal activities are published very often in national and regional newspapers. Considering its importance in various fronts, it is imperative to review its socio-economic status, trade and commerce, and these are reviewed in this chapter.

14.2 Vulnerability, Smuggling and Illegal Activities

Several dynamic forces are responsible for the vulnerable status of red sanders which pose challenges to the government, conservationists, ecologists, traders, cultivators and tribals residing in this sensitive zone. For example, even minor disturbances like cattle grazing and forest fires cause severe damage to the seedlings of red sanders (Ramakrishna 1962). Yet another threat to the red sanders population is due to fuelwood collection and overgrazing by cattle. The tree is very sensitive at seedling stage, and hence, growth is hampered due to grazing and fire. The endemic nature of the tree, slow growth, distribution patterns, prolonged juvenile period on the one hand and the international demand for the quality of wood on the other hand resulted in over-exploitation of the tree. Red sanders was classified as endangered species in the *1997 IUCN Red List of Threatened Plants*, and the government of India proposed its inclusion in CITES Appendix II (Arunkumar 2011; Soundararajan and Joshi 2012). It is also cited as endangered in *The World List of Threatened Trees* (Oldfield et al. 1998). Adding to this restricted trade and ban on the felling of the tree attracted illegal felling. Ramakrishna (1962) mentioned that commercial exploitation of red sanders is restricted in 1956, and the working plan of Rajampet presented in 2008 reiterates that it is stopped during 1980 (Kukrety 2011). This resulted in huge demand supply gap especially in the international market. Smuggling is rampant, and this is a cause of concern, and day by day the situation is getting worsened and hence vulnerably pushing it to the threatened status. Kukrety (2011) deliberated CITES report (2007) and provides data that there is seizure of 947 tonnes of red sanders between 2003 and 2005. Further he states that the supply of illegal wood happens through Nepal, Myanmar, Singapore and Taiwan (non-source countries) before it reaches its final destinations. Andhra Pradesh Forest Department (2010) reported seizure of 3067 tonnes of wood between 2001 and 2007. This is the only way the seized wood enters into the legal market through open auction. Mulliken and Crofton (2008) made a mention of estimated demand of 3000 tonnes which is far lesser than the actual supply. The commercial exploitation in the past without adequate replenishment can be censured for the descent of natural strands of red sanders. If similar trends continue in the future, fragmentation and subsequent extinction may be impending.

There is an inevitable demand supply gap because of varied reasons, and this propelled smuggling and illegal activities. There is a mention of these activities in the early 1960s itself by Ramakrishna (1962), and the evidence of seizers and criminal cases kept on progressing even today. CITES (2007) reports indicated seizure of 947 tonnes of wood between 2003 and 2005, and APFD (2010) states smuggling up to 3076 tonnes between 2001 and 2007 (Bhagyaraj and Ramana 2013). An excellent review about the number of cases booked and quantity of wood held by the Andhra Pradesh Government is published by Bhagyaraj (2017). According to him a total of 2758 cases were booked from 2003 to 2016. The data is alarming, and the number of persons arrested is 9000, and the total quantity of wood seized is 9913 metric tonnes which really matters and causes a concern. Of these, 92% of seizures happened in Chittoor district alone. Every day there are news headlines pertaining to seizures and criminal activities happening, and the situation is appalling (Table 14.1).

14.3 Restoration

Forest departments of respective states are very serious, and there were attempts to restore and retain the resilience of the degraded ecosystem of red sanders. Though the tree belongs to certain forest tracts of southern plateau of India, there are unsung efforts by international agencies like Rufford Foundation. In unassailable task of creating awareness about the tree, importance of plantation, distribution of saplings, and efforts to conserve the tree by roping in local communities are taken up during 2007–2008 by M. Anuradha, and the project is supported by Rufford Foundation, United Kingdom (www.rufford.org/rsg/projects/anuradha_maniyam). However, there is a limited research and publications about its ecology, sustainable utilization, repeatability and scaling up of the standardized cultivation practices. These are the grey areas and need immediate attention from every stakeholder to restore the precious tree. Restoration of red sanders landscape involves protection of habitats, populations, related species conservation, ecological interaction, community involvement, rigorous plantations, in situ conservation and most importantly constitution and implementation of comprehensive policy.

Table 14.1 News coverage about red sanders illegal felling and smuggling

S. no.	Title	References
1.	DRI seizes red sanders worth Rs. 9 crore	Times of India (2018)
2.	Andhra Pradesh engineer among three red sanders smugglers held, 22 logs seized	New Indian Express (2018)
3.	Rs. 30 crore red sanders smuggling: government may reinvestigate the case	Times of India (2018)
4.	Encounters no solution to red sanders smuggling problem: PCCF	The Hindu (2018)
5.	218 red sanders logs seized, one held	New Indian Express (2018)
6.	Dog squad helps officials in nabbing red sanders woodcutter	The Hindu (2018)
7.	Chennai: two cops, home guard attacked during probe into red sanders smuggling	New Indian Express (2018)
8.	Red sanders wood worth about Rs. 4 crore seized in Tirupati	Business standard (2018)
9.	Cops raid Bengaluru godown, seize red sanders logs	The Hindu (2018)
10.	Gang gags guard, loots red Sanders worth lakhs	Times of India (2018)
11.	Andhra Pradesh: 140 more cops deployed to curb smuggling of red sanders logs	New Indian Express (2018)
12.	Probe ordered into "robbery" of red sanders from godown	The Hindu (2018)
13.	Red sanders logs seized after hot chase	The Hindu (2018)
14.	99 red sanders logs worth Rs. 2.50 crore seized	The Hindu (2018)
15.	Seven red sanders seized	New Indian Express (2018)
16.	Forest beat officer, 16 others held for smuggling red wood in Nellore	New Indian Express (2018)
17.	Jawadhu Hills man held with red sanders	Times of India (2018)
18.	Forest officials seize red sanders worth Rs. 3 crore in Kadapa	Uni India (2018)
19.	Andhra Pradesh: red sanders protection plan in disarray reveals forest meeting	Indian Express (2018)
20.	After the crackdown: the red sanders smuggler's new, longer route from Andhra to Chennai port	Indian Express (2018)
21.	International red sanders smuggler, kin, aides held	The Hindu (2018)
22.	Forest department to auction over 80 MT of seized red sanders wood today	Times of India (2018)
23.	Interstate red sanders smuggler held in Tamil Nadu	The Hindu (2018)
24.	Red sanders logs found in car	The Hindu (2018)
25.	Why red gold is too hot for locals to touch	Economic Times (2017)
26.	Smugglers adopt new tactics for transporting red sanders logs	The Hindu (2018)
27.	Twelve tonnes of red sanders seized (29/03/2018)	The Hindu (2018)
28.	Red sanders logs seized	The Hindu (2018)
29.	10.1 metric tonnes red Sanders recovered from China-bound vessel in Visakhapatnam	New Indian Express (2018)

(continued)

Table 14.1 (continued)

S. no.	Title	References
30.	Anantapur: special teams nab key red sanders smuggler	Deccan Chronicle (2018)
31.	Thoothukudi: Rs. 6 crore worth red sanders in customs custody go missing	Deccan chronicle (2018)
32.	26 red sanders logs seized, seven smugglers arrested in Andhra Pradesh	New Indian Express (2018)
33.	Unemployed youth of TN fall prey to red sanders smugglers	The Hindu (2018)
34.	AP police nab 84 from TN allegedly heading to cut red sanders trees	The Hindu (2018)
35.	5 bodies afloat in Andhra Pradesh lake; locals say red sanders looters	Times of India (2018)
36.	26-year-old woman and four others arrested for smuggling red sanders	Times of India (2018)
37.	Nellore: red sanders logs seized, 9 smugglers held in raid	Deccan Chronicle (2018)
38.	Rs. 3 crore red sanders seized, kingpin's brother arrested	Times of India (2018)
39.	Seven red sanders logs seized, 15 TN workers arrested	New Indian Express (2017)
40.	Three tones of red sanders logs seized	The Hindu (2017)
41.	Red sanders worth Rs. 16 crore seized in raid	Times of India (2017)
42.	13 TN men arrested by AP red sanders antismuggling task force	Times of India (2017)
43.	Most-wanted red sanders smuggler, 10 others held	The Hindu (2017)
44.	Sleuths stumble upon cave with red sanders	The Hindu (2017)
45.	Interstate red sanders smugglers held in raid	New Indian Express (2017)
46.	International red sanders smuggler, aides held	The Hindu (2017)
47.	158 logs, list of red sanders smugglers seized in forest	New Indian Express (2017)
48.	Red sanders seized, 6 held	The Hindu (2017)
49.	Rival gangs fight over red sanders, get caught by police	The Hindu (2017)
50.	Red sanders logs recovered from Tamil Nadu godown	The Hindu (2017)

14.4 Trade and Commerce

Red sanders is one of the main lucrative markets for smugglers and traders as a high price is paid for this wood in the export market, and there are growing trends of underground market and arrests of the red sanders smugglers (Bhagyaraj 2017). The earliest historical record was evidenced in China documented by Cui Bao of Jin Dynasty (265–340 A.D.) in "Comments on Ancient and Modern Items". In this the wood is regarded as a rare and precious lignum as expensive as gold. With respect to trade, this can be traced back to 1681 when the British East India Company transported to England for dyeing purpose (Reddy 1972).

The trade and export of red sanders dates to the seventeenth century and has its reference during Ming and Qing periods by Chinese. Tzu-t'an in Chinese is etymologically linked to the Indian name *Rakta chandan*, and Chinese traders are familiar with red sanders (Bhagyaraj 2017). Export of red sanders from India to Europe commenced in the seventeenth century mainly for dyeing the textile, and there was also demand for wood, and reportedly exports tuned up to 3000 tonnes per annum on average. The wood was exported to Europe and other countries, and Ramakrishna (1962) made a mention of Gamble's statement regarding several tonnes of export to the United Kingdom and France. The demand for the wood in railways also caused a serious threat; between the years 1870–1871 and 1881–1882, around 150,472 tonnes of wood was supplied to the railways and almost from the red sanders forests of Rajampet. However, this found to cease due to the competition from synthetic dyes. Again, the export gained its pace when the ripple-grained wood considered to have superior acoustic qualities, which has a huge demand in Japan. The wavy-grained wood is used for making a musical instrument called as "*Shamisen*", a three-stringed instrument, and in 1930, Japan started importing; however, due to over-exploitation, the trading of the wild resources was controlled during 1950 and 1960 (Ramabrahmam and Sujatha 2016; Bhagyaraj 2017). The wood extract has its usage as a colourant in fish processing and in food industry by the Europeans, and the export of red sanders powder from India averaged to 50 tonnes between 1988 and 1993, and the major importers are Japan and China (Taiwan province) and Europe (Bhagyaraj 2017). Soundararajan and Joshi (2012) reviewed grading, rules and regulations for export of red sanders. Bhagyaraj (2017) highlighted the interventions of the Andhra Pradesh Government to regulate and protect red sanders.

14.5 Socio-economic Impact of Red Sanders

Though there is a lucrative market for red sanders, there is no encouraging response for the farmers to take up cultivation as this is a very slow-growing tree, and the harvestable size is attained only after 25 years of age. The heavy demand in international market for many years is the reason for over-exploitation and illegal harvest. In Andhra Pradesh alone during 2012–2013, 1488 forest offence cases were booked for smuggling red sanders (Arunkumar and Joshi 2014); 1002 vehicles carrying 1390 tons of wood is reported to be seized (Anonymous 2014). Devious smugglers are taking the advantage of the penury of the forest dwellers, encompass villagers and entice them for felling the mature red sanders trees from the forests (Anonymous 2014).

The socio-economic status, innocence and ignorance of the communities surrounding the red sanders ecosystem played a pivotal role and may be a reason for getting tempted for the lures of illegal traders. The newspaper articles say thousands of poor labourers are transported from Tamil Nadu to Andhra Pradesh and involved in the illegal felling which is done in the disguise of wood collectors.

The men involved in this entire operation are innocent and just lured for the wages. This is termed as well-oiled network, and the racket operates from tribal villagers to international markets (Hindustan Times 2015). There are many other critical factors like cattle grazing and villagers setting fire to the grass to secure fresh flush of grasses for the grazing (Kukrety 2011). The local demand of red sanders is not much as compared to international demand; hence, there is an involvement of international traders which will not fetch the local communities and smallholders, and these people cannot expect lucrative benefits. Sayer et al. (2004) opined that unless there are direct economic benefits and indirect incentives like environmental and social services, the efforts to restore are unlikely sustainable. As repeatedly reiterated the red sanders market is driven by illegal felling; this is having a negative impact societally even creating threats to the life of innocent labourers. On the other side, bans on collection, stringent restrictions, administrative procedures of the Andhra Pradesh Forest Department and modus operandi as per CITES for harvest and marketing are lengthy and unwieldy, discouraging private landowners to raise red sanders plantations. This even forces the smallholders to sell the trees for throwaway price to the middle man and has an impact on the socio-economic conditions of local communities.

14.6 Challenges and Perspectives for Stakeholders

Frequent changing of smuggling routes is a challenge and needs constant vigilance and efforts from all the neighbouring countries (Arunkumar and Joshi 2014). There are frequent smuggling cases and illegal trafficking is reported, and this is a news item almost every month even now. Considering its illegal trade, harvest and destruction in its habitat, the species is categorized as endangered by the IUCN and listed in the Convention on International Trade in Endangered Species of Wild Fauna and Flora (CITES), an organization which ensures survival of wild plants and animals traded internationally and safeguards the species from over-exploitation. It is protected under red sanders and Sandalwood transit rules of Andhra Pradesh Forest Act, 1967, and is also classified as a reserved tree under the Andhra Pradesh Preservation of Private Forest Rules, 1978 (Arunkumar and Joshi 2014). Augmented peril to the red sanders forests impelled the government of India and the government of Andhra Pradesh to restrict commercial harvest in 1956 (Ramakrishna 1962), and a total ban on harvest in 1980 was presented in the working plan Rajampet (2008). As discussed earlier the red sanders is included in the CITES based on the recommendations from India in 1995 which states that chips, powder and timber of red sanders are not permitted for trade, as it is classified in Appendix II of CITES as endangered – EN B1+2de with threats involving a combination of overharvest and habitat alteration (IUCN 2010). This further extends to the need for clearance from both state and central governments even for trading extracts. The most important stakeholder worth mentioning is a farmer or cultivator. Though the market is assured indicating high demand, the species is not opted for plantation by private planters. During the survey conducted by Dr. Anuradha, who worked in a project sponsored by Rufford Foundation, it was noted that very few growers planted on bunds, along the fence interspersed with other timber-yielding plants. This is due to lack of information on practices, harvest period, market data and delay in administrative clearances. Teak is better adopted by local planters due to extensive extension programmes conducted over a period. The Andhra Pradesh Forest Department though taken up planting in 4099 ha during 1960–1975 in Chittoor, Kadapa and Kurnool districts, commercial exploitation of the same is not taken up (APFD 2010). Many times, there is an opinion that the wood produced in commercial plantations is often considered to be inferior when compared to the wild variety, whereas Indu and Anuradha (2018) felt that there are superior trees even in the private lands and plantations, where they have collected the pencil thick wood by using the increment borer. The survey and data collected from the plantations of Jarakabande and Karnataka revealed that there are mixed populations of plus trees and inferior trees. The samples collected by using increment borer are subjected to santalin content analysis and measurements such as height, girth and straight bole. These plantations are uniformly aged-grown under similar environmental conditions, however exhibited mixed characters. This clearly indicates that the santalin content and superior phenotype characters are more genetic rather than environment driven.

14.7 Conclusion

Red sanders, the pride of Andhra Pradesh, India, is facing enormous threat, and the reasons are reiterated and published extensively. Though this is discussed and deliberated by national and international agencies, there is lack of comprehensive and combined efforts. All the stakeholders starting from tribal to exporter should join their hands to bring out a strategy for the conservation and sustainable utilization of red sanders.

References

Anonymous (2014) Red alert on red sanders. Sri Lanka apprehends large consignment from India. Traffic Post 20:8–9

APDF (2010) Andhra Pradesh Forest Department, inventory report. http://www.forests.ap.gov.in/abkp/pdf/forest_inventory_report.pdf. Accessed on 8/07/18

Arunkumar AN (2011) Variability studies in *Pterocarpus santalinus* in different aged plantations of a Karnataka. Myforest 47(4):343–353

Arunkumar AN, Joshi G (2014) *Pterocarpus santalinus* (red sanders) an endemic, endangered tree of Indian: status, improvement and the future. J Trop For Environ 4(02):1–10

Bhagyaraj A (2017) Red sanders invisible activities in Chittoor district: case study. Int J Adv Res 5 (9):1073–1080

Bhagyaraj A, Ramana DV (2013) Status of red sanders in business. Indian J Appl Res 3(7):405–407

CITES (2007) Review of significant trade in specimens of appendix II species-PC17 Doc. 8.4. http://www.cites.org/eng/com/PC/17/E-PC17-08-04.pdf. Accessed 08/07/18 http://www.newindianexpress.com/cities/chennai/2018/jun/23/chennai-two-cops-home-guard-attacked-during-probe-into-red-sanders-smuggling-1832324.htm. Accessed on 26/06/2018

http://www.newindianexpress.com/states/andhra-pradesh/2017/dec/08/158-logs-list-of-red-sanders-smugglers-seized-in-forest-1721697.html. Accessed on 28/06/2018

http://www.newindianexpress.com/states/andhra-pradesh/2017/dec/21/seven-red-sanders-logs-seized-15-tn-workers-arrested-1733098.html. Accessed on 27/06/2018

http://www.newindianexpress.com/states/andhra-pradesh/2017/nov/17/inter-state-red-sanders-smugglers-held-in-raid-1703179.html. Accessed on 28/06/2018

http://www.newindianexpress.com/states/andhra-pradesh/2018/jun/15/andhra-pradesh-140-more-cops-deployed-to-curb-smuggling-of-red-sanders-logs-1828501.html. Accessed on 26/06/2018

http://www.newindianexpress.com/states/andhra-pradesh/2018/jun/20/218-red-sanders-logs-seized-one-held-1830845.html. Accessed on 26/06/2018

http://www.newindianexpress.com/states/andhra-pradesh/2018/jun/21/forest-beat-officer-16-others-held-for-smuggling-red-wood-in-nellore-1831362.html. Accessed on 26/06/2018

http://www.newindianexpress.com/states/andhra-pradesh/2018/jun/26/andhra-pradesh-engineer-among-three-red-sanders-smugglers-held-22-logs-seized-1833665.html. Accessed on 26/06/2018

http://www.newindianexpress.com/states/andhra-pradesh/2018/mar/09/101-metric-tonnes-red-sanders-recovered-from-china-bound-vessel-in-visakhapatnam-1784490.html. Accessed on 27/06/2018

http://www.newindianexpress.com/states/andhra-pradesh/2018/mar/18/26-red-sanders-logs-seized-seven-smugglers-arrested-in-andhra-pradesh-1788930.html. Accessed on 27/06/2018

http://www.newindianexpress.com/states/andhra-pradesh/2018/may/28/seven-red-sanders-seized-1820294.html. Accessed on 26/06/2018

http://www.thehindu.com/news/national/andhra-pradesh/80-tn-red-sanders-woodcutters-arrested-in-ap/article22905185.ece. Accessed on 27/06/2018

http://www.thehindu.com/news/national/andhra-pradesh/cops-raid-bengaluru-godown-seize-red-sanders-logs/article24030183.ece. Accessed on 26/06/2018

http://www.thehindu.com/news/national/andhra-pradesh/dog-squad-helps-officials-in-nabbing-red-sanders-woodcutter/article24235681.ece. Accessed on 26/06/2018

http://www.thehindu.com/news/national/andhra-pradesh/encounters-no-solution-to-red-sanders-smuggling-problem-pccf/article24117617.ece. Accessed on 26/06/2018

http://www.thehindu.com/news/national/andhra-pradesh/international-red-sanders-smuggler-kin-aides-held/article23806708.ece. Accessed on 27/06/2018

http://www.thehindu.com/news/national/andhra-pradesh/inter-state-red-sanders-smuggler-held-in-tamil-nadu/article23517426.ece. Accessed on 27/06/2018

http://www.thehindu.com/news/national/andhra-pradesh/red-sanders-logs-seized/article22683570.ece. Accessed on 27/06/2018

http://www.thehindu.com/news/national/andhra-pradesh/smugglers-adopt-new-tactics-for-transporting-red-sanders-logs/article22778911.ece. Accessed on 27/06/2018

http://www.thehindu.com/news/national/andhra-pradesh/unemployed-youth-of-tn-fall-prey-to-red-sanders-smugglers/article23809359.ece. Accessed on 27/06/2018

http://www.thehindu.com/news/national/tamil-nadu/red-sanders-logs-found-in-car/article22954087.ece. Accessed on 27/06/2018

http://www.thehindu.com/news/national/tamil-nadu/three-tonnes-of-red-sanders-logs-seized/article21011543.ece. Accessed on 27/06/2018

http://www.thehindu.com/news/national/tamil-nadu/twelve-tonnes-of-red-sanders-seized-thoothukudi/article23383643.ece. Accessed on 27/06/2018

http://www.thehindu.com/todays-paper/tp-national/tp-andhrapradesh/99-red-sanders-logs-worth-rs-250-cr-seized/article24141138.ece. Accessed on 26/06/2018

https://economictimes.indiatimes.com/news/environment/flora-fauna/why-red-gold-is-too-hot-for-locals-to-touch/articleshow/60816699.cms. Accessed on 27/06/2018 https://indianexpress.com/article/india/after-the-crackdown-the-red-sanders-smugglers-new-longer-route-from-andhra-to-chennai-port-5079518/. Accessed on 27/06/2018

https://indianexpress.com/article/india/andhra-pradesh-red-sanders-protection-plan-in-disarray-reveals-forest-meeting-5127180/. Accessed on 27/06/2018

https://timesofindia.indiatimes.com/city/ahmedabad/dri-seizes-red-sanders-worth-rs-9cr/articleshow/64725894.cms. Accessed on 26/06/2018

https://timesofindia.indiatimes.com/city/amaravati/5-bodies-afloat-in-andhra-pradesh-lake-locals-say-red-sanders-looters/articleshow/62976059.cms. Accessed on 27/06/2018

https://timesofindia.indiatimes.com/city/chennai/26-year-old-woman-and-four-others-arrested-for-smuggling-red-sanders/articleshow/63261032.cms. Accessed on 27/06/2018

https://timesofindia.indiatimes.com/city/chennai/jawadhu-hills-man-held-with-red-sanders/articleshow/64567809.cms. Accessed on 26/06/2018

https://timesofindia.indiatimes.com/city/chennai/red-sanders-worth-rs16-crore-seized-in-raid/articleshow/61716961.cms. Accessed on 28/06/2018

https://timesofindia.indiatimes.com/city/chennai/rs-30-crore-red-sanders-smuggling-government-may-reinvestigate-case/articleshow/64739818.cms. Accessed on 26/06/2018

https://timesofindia.indiatimes.com/city/mumbai/rs-3-crore-red-sanders-seized-kingpins-brother-arrested/articleshow/63209841.cms. Accessed on 27/06/2018

https://timesofindia.indiatimes.com/city/puducherry/gang-gags-guard-loots-red-sanders-worth-lakhs/articleshow/64375178.cms

https://timesofindia.indiatimes.com/city/thane/forest-dept-to-auction-over-80-mt-of-seized-red-sanders-wood-today/articleshow/64180814.cms. Accessed on 27/06/2018

https://timesofindia.indiatimes.com/india/ap-red-sanders-anti-smuggling-task-force-arrests-20-tamil-speaking-men-in-tirumala/articleshow/63959954.cms. Accessed on 28/06/2018

https://www.business-standard.com/article/pti-stories/red-sanders-wood-worth-about-rs-4-cr-seized-in-tirupati-118061901159_1.html. Accessed on 26/06/2018

https://www.deccanchronicle.com/nation/crime/030318/thoothukudi-rs-6-crore-worth-red-sanders-in-customs-custody-go-missin.html. Accessed on 27/06/2018

https://www.deccanchronicle.com/nation/crime/160218/anantapur-special-teams-nab-key-red-sanders-smuggler.html. Accessed on 27/06/2018

https://www.deccanchronicle.com/nation/crime/240318/nellore-red-sanders-logs-seized-9-smugglers-held-in-raid.html. Accessed on 27/06/2018

https://www.hindustantimes.com/india/red-sanders-smuggling-a-well-oiled-network-Hindustantimes. Accessed on 20/06/2018

https://www.rufford.org/rsg/projects/anuradha_maniyam. Accessed on 25.06.2018

https://www.thehindu.com/news/cities/bangalore/rival-gangs-fight-over-red-sanders-get-caught-by-police/article20234667.ece. Accessed on 28/06/2018

https://www.thehindu.com/news/national/andhra-pradesh/most-wanted-red-sanders-smuggler-10-others-held/article19958157.ece. Accessed on 28/06/2018

https://www.thehindu.com/news/national/andhra-pradesh/red-sanders-logs-recovered-from-tamil-nadu-godown/article20493632.ece. Accessed on 28/06/2018

https://www.thehindu.com/news/national/andhra-pradesh/red-sanders-seized-6-held/article19863542.ece. Accessed on 28/06/2018

https://www.thehindu.com/news/national/andhra-pradesh/sleuths-stumble-upon-cave-with-red-sanders/article22467289.ece. Accessed on 28/06/2018

https://www.thehindu.com/news/national/tamil-nadu/international-red-sanders-smuggler-aides-held/article19556024.ece. Accessed on 28/06/2018

Indu BK, Anuradha M (2018) Optimization of marker assisted technology for screening, selection and micro propagation of elite genotypes of *Pterocarpus santalinus* (red sandalwood). http://online-wosa.gov.in:8080/wosa/wos/project?

IUCN (2010) International union for conservation of nature – red list of threatened species. http://www.iucnredlist.org/apps/redlist/details/32104/0. Accessed 08/07/18

Kukrety S (2011) Restoration of red sanders (*Pterocarpus santalinus* L.) forests for ecological and economic benefits. PhD thesis, University of Florida
Mulliken T, Crofton P (2008) Review of the status, harvest, trade and management of seven Asian cites-listed medicinal and aromatic plant species. https://www.cites.org/sites/default/files/common/com/pc/17/X-PC17-Inf-10.pdf. Accessed 08/07/18
Oldfield SO, Lusty C, MacKinven A (1998) The world list of threatened trees. World Conservation Press, Cambridge, p 650
Ramabrahmam DV, Sujatha (2016) Red sanders in Rayalaseema Region of Andhra Pradesh; importance to commercial and medicinal value. IOSR J Pharm Biol Sci 11(IV):57–60
Ramakrishna A (1962) The red sanders and its future. Indian For 3:202–205
Reddy CVK. (1972) Red sanders and its history of utilization. Indian For 589–93
Sayer J, Chokkalingam U, Poulsen J (2004) The restoration of forest biodiversity and ecological values. For Ecol Manag 201:3–11
Soundararajan V, Joshi SC (2012) Grading, rules and regulations. In: Endemic possessions of Eastern Ghats: red sanders (*Pterocarpus santalinus* Linn. f). Institute of Wood Science and Technology (ICFRE), Bangalore, pp 80–95
Working plan Rajampet (2008) Working plan for the management of forests of Rajampet division. Andhra Pradesh Forest Department, Government of Andhra Pradesh India

Tree Improvement in Red Sanders

15

B. K. Indu, R. Kavyashree, S. Balasubramanya, and M. Anuradha

Abstract

Red sanders (*Pterocarpus santalinus*) is acclaimed for its fabulous timber character and is highly adored for its red-coloured dye santalin and opulent heartwood. This highly prized endangered and endemic tree lured illicit international trade and is controlled by ruthless mafia. The tree has typically two different qualities which determine their value, one has wavy grain wood texture impregnated with intense scarlet red santalin principles and the other with straight grained texture with relatively light colour. Because of this mixed population of elite and non-elite genotypes, red sanders deserves tree improvement programs for its sustainable utilization. Any tree improvement program can be successful with the availability of information on phenology, reproduction biology, genetic and molecular status and breeding techniques. In this review various methods practiced for improvement, limitations and future scope is discussed.

Keywords

Red sanders · Tree improvement · Phenology · Santalin · Plus tree selection

B. K. Indu (✉)
Rishi Foundation, Bangalore, India

R. Kavyashree
Department of Biotechnology, The Oxford College of Science, Bangalore, Karnataka, India

S. Balasubramanya
Department of IT, BT and S&T, Government of Karnataka, Karnataka Innovation & Technology Society, Bangalore, India

M. Anuradha
Department of Biotechnology, Padmashree Institute of Management and Sciences, Bangalore, Karnataka, India

T. Pullaiah et al. (eds.), *Red Sanders: Silviculture and Conservation*,
https://doi.org/10.1007/978-981-13-7627-6_15

15.1 Introduction

Forest tree breeding is gaining its momentum worldwide, and industry has invested in such programs with great expectations (Kedharnath 1984). The author has aptly mentioned that these programs helped in the intensive forest management through artificial regeneration and establishment of plantations. Due to focus on improvement, forests are increasingly restored, and this offered a great opportunity for better management and helped in choosing appropriate genotypes. Tree improvement process is basically designed to enhance the quality in terms of physical and chemical properties responsible for adoption to challenging environments and suitable for the commercial applications. Various parameters to be considered for the success of such programs are enhancing wood quality, yield, productivity, resistance to pests, diseases and adoptability to extreme environmental conditions.

Success in tree improvement programs can lead to improved yield and help in rapid increase of commercial plantations that in turn will have a positive bearing on the conservation of genetic resources and diversity. Designing and development of efficient improvement strategy requires genetic knowledge pertaining to the breeding population, its selection methods, criteria and sizes (Comstock 1977). Most importantly the basic information of phenology, reproductive biology, propagation and molecular genetics is imperative.

Tree breeding and improvement programs are hampered because of their prolonged juvenile period, complex phenology, widespread distribution and lesser understanding of wide range of morphological and genetic variations. Tree improvement programs emphasize on the exploitation of this variation (Puri 1998) and accordingly set out to isolate and evaluate the genetic variation in one or more traits of interest. These variations existing in tree species provides an opportunity to develop improved varieties. Variations are not only determined by its genes but also by epi genes which depends on the environment in which trees are growing. Red Sanders known for its characteristic timber colouring principles and medicinal value displays a wide range of variability at morphological, biochemical and molecular level. From tree improvement perspective, documenting variability in a tree species is a prerequisite before initiating any tree improvement program. Many researchers have reported variability for different traits in red sanders (Padmalatha and Prasad 2007; Arunkumar 2011; Rani and Usha 2013; Jyothi et al. 2014).

15.2 Tree Improvement Strategies

Red sanders has many unique features both in physical and chemical properties, and the variations are distributed across the population. Due to various reasons, the attempts to develop improved varieties are limited and yet to gain momentum. The factors and strategies with limitations are critically reviewed.

The broad strategies which need to be adopted for tree improvement program are conventional propagation methods, breeding techniques and biotechnological approaches. The conventional approach of improvement is through selection, testing/breeding and evaluation. Tree breeding concept is relatively simple wherein potential 'parent' trees are selected from large number of candidates, and their performance is evaluated by growing their progeny. The parent trees with best performance are selected for improvement programs. This approach results in establishment of plus-tree progeny leading to improvement in quality and yield (Puri 1998).

The first and foremost requirement is selection of plus trees with desired characters. In tree improvement programs, the focus is on its timber quality, followed by resistance to pests and other environmental factors. Currently biotechnologists are strategically designing methods to reduce the lignin content in lieu of disadvantage of high lignin in paper industry. In case of red sanders, the main strategy is to secure wavy grained timber with high santalin content apart from achieving more girth and straight bole which is a general feature that determines quality of the timber. Either of the approach needs to get initiated with fixing the desired characters and selection of the same. The selected trees with elite characters are being utilized in tree improvement programs.

15.3 Selection of Plus Trees

As discussed, this is the initial step in any tree improvement program which enables the manipulation of variability in a population in the desired direction. The fact that genes vary in their expression and that individuals inherit different combination of these genes provides the basis for selection. Through selection a progressively improved breeding population is created. This is achieved through series of selection cycles in which genetic variabilities of economic traits are measured, manipulated and delivered to improve the traits over time (Puri 1998).

The major challenges are to identify precisely the genes responsible for manifesting desired characters, and prolonged juvenile period makes tree improvement programs difficult. However, one technique that overcomes this challenge is plus-tree selection, which includes selection of superior trees (plus trees) in the best forest stands. Superior or Plus tree is one that has superior phenotype for growth, form, narrow crown, resistance to pests and diseases, wood quality or other desired characteristics. Plus tree is an outstanding tree that occurs in natural stands having as many desirable traits as possible. However, locating such trees in the wild populations requires keen observation and expertise as such plus trees occur in very low frequency. There are many attempts by various researchers to relate certain morphological and other parameters to the elite characters of red sanders (Kedharnath and Rawat 1976; Lohidas and Dayanand 1984; Rawat and Uniyal 1996).

Various approaches are practiced for plus-tree selection, and it depends on the type of genetic variation present in the population and the type of stands available. One approach is 'tandem selection' wherein one character at a time is focused over successive breeding generations and hence not preferred for trees due to their long gestation period. Another approach is 'comparison-tree selection' wherein a 'candidate' tree is compared with number of neighbouring trees for various character or traits. For each character or trait there is a fixed level of merit and the candidate tree must reach this level for acceptance. The candidate tree is rejected if it fails to reach this fixed level despite of its acceptability in other characters. However, in this approach care should be taken to choose trees with certain minimum distance for comparison to avoid genetic relatedness (Ledig 1974). The third approach involves 'index' wherein the character or trait is weighed in a systematic manner considering relative economic value of each character, phenotypic and genotypic variations and covariance of each pair of character. The last approach is the 'base line selection' that is used when comparison with neighbouring trees is hampered due to uneven-aged or sized strands. In this approach, a candidate tree is first identified for their overall superiority and then certain traits are chosen and compared with five dominant check trees in the vicinity. Grading system is followed wherein each character is assigned a score ranging from 0 to 5 with 5 as the most desirable expression. If the total score exceeds a given minimum, then the tree is accepted as a plus tree (Puri 1998). Of the four approaches discussed above, the second approach 'comparison-tree selection' is the most commonly used selection technique.

In red sanders the superior trees are with wavy grain heartwood impeded with deep red-coloured dye santalin apart from other features such as straight stem, narrow crown and less branching. The desired variant with rippled grain and high santalin content seems to occur in low frequency in nature, maybe one in hundred. They seem to show no apparent morphological differences by which they can be easily recognized from the normal grained trees (Kedharnath 1984). Moreover, the grain nature is clearly differentiated only after attaining maturity; as it is invasive and threats the life of tree, selection based on the heartwood characters is becoming very difficult. However, the increment borer technique can be used for extracting pencil thickness wood at breast height without causing damage to the tree. Authors Indu and Anuradha collected the samples at Jarakabande plantation in Karnataka. As per the observation tall trees with more girth, straight bole, is invariably has high santalin content and more heartwood to sap wood ratio (unpublished). Based on these methods, plus trees can be selected for either conventional or applied tree improvement programs.

Often *P. santalinus* is described as 'red gold'; tree improvement activities have been carried out on this species during which 25 candidate plus trees have been identified by the Forest Department of Andhra Pradesh (Reddy and Srivasuki 1992).

15.4 Reproductive Biology

Complete understanding of reproductive biology is an essential prerequisite for tree improvement programs. The pollination ecology is less investigated in red sanders. Rao and Raju (2002) studied phenology and reproduction biology of red sanders and felt that there is lower natural fruit set when compared to flower production. The comprehensive study states that *Pterocarpus santalinus* is able to initiate fruit production both by self- and cross-pollination; however selective elimination of poor offspring resulting from self-pollination is gradually eliminated. Further the variations in this species are expected to be lower because of its endemic status in turn reduced opportunity for out-crossing and enforcing inbreeding. Due to dry and hot habitat, there is less or nil pollinator activity during day time, and hence self-compatibility through geitonogamy is inexorable for red sanders. Rao and Raju (2002) revealed that in *P. santalinus* also though geitonogamy exists, still it mainly cross-pollinates and exhibits facultative xenogamous breeding system and this is considered as advantageous ensuring continued survival. Tree breeding itself is time-consuming and less practiced. However, understanding reproductive biology will help to integrate research of conventional breeding and biotechnological tools like anther culture and mutation studies.

In tree breeding programs, the shoots from the superior or plus trees are grafted onto seedling rootstock in the nursery. The flowers produced from these grafts are subjected to controlled crossing for seed production to ensure best traits. The seeds thus produced are collected and sown to produce offsprings, which are then planted out in progeny tests on typical forest sites. Only those plus trees that produce superior progeny are used as breeding stock for the next generation. It can take approximately 10 years to establish reliable information on the performance of the progeny.

In the case of red sanders, the germination and establishment of seedlings itself is a challenge and resolved to some extent by various researchers (Dayanand and Lohidas 1988; Naidu 2000, 2001a, b; Naidu and Rajendrudu 2001; Kalimuthu and Lakshmanan 1995; Arockiasamy et al. 2000; Anuradha and Pullaiah 1998). There is no research done to establish protocols for emasculation and inbreeding in red sanders. The other alternative is to achieve this by vegetative propagation and grafting.

15.5 Vegetative Propagation

Forest department has a mandate for undertaking trials and establishment of plantations. The areas prioritized are identification of superior trees, seed production, storage, testing, seed treatments, germination studies, testing the purity of parental characters, clonal multiplication, raising plantations and field performance at different geographical zones. Vegetative propagation through rooting the cuttings is a herculean task, and this tree is categorized under difficult-to-root species. Reddy and Srivasuki (1990) tried to root semi hard wood cuttings, cleft grafting and air layering with limited success. They later improvised the percent of rooting by modifying the treatments during the process of air layering (Reddy and Srivasuki 1992).

In red sanders, the stumps of the trees coppice profusely. These coppices can be rooted, and the technology can be adopted for establishing large number of clones. As the stumps provide an opportunity to visibly estimate the quality of the heartwood, girth of the tree and santalin content, the selection of the superior trees is also possible. Wedge grafting to ordinary stock with quality scions of wavy grain sounds to be a successful strategy but has limited scope because of high mortality and incompatibility (Soundararajan and Joshi 2012). Optimization of a successful clonal multiplication technology can ease the entire process of tree improvement program and establishment of large number of superior planting stocks.

Seeds or vegetative material/grafts of those plus trees with the best breeding value are used to create seed orchards for the production of improved planting stock. Though seed propagation is one of the simplest ways to mass propagate, it produces genetically heterozygous population with variable growth characteristics. This may lead to an undesirable outcome for forestry wherein clonal, genetically uniform material is preferred (Teixeira da Silva et al. 2018). Hence, if the species is easy to mass-produce vegetatively, then clone banks can be established for production of cuttings. The time scale for the selection and testing phases for species through evaluation of progeny trials will be at least 8–20 years based on the species being improved and the level of improvement being undertaken. Establishment of seed or clonal seed orchards will usually take 1–3 years and is dictated by the time required to collect the seeds or cuttings and raise the plants. Production of seeds from the seedling plantations may take at least 8–12 years, which mainly depends on the time of flowering. In case of clonal seed orchards where vegetative material is used seed, production can be much earlier. As discussed earlier clonal propagation by vegetative means is achieved with a limited degree of success in red sanders; hence the other alternative is to adopt biotechnological approaches.

15.6 Biotechnological Strategies

Tree improvement by conventional methods such as identification, selection of superior trees and breeding programs involving the superior trees is laborious and highly time-consuming. The major limiting factor for tree improvement is prolonged juvenile period, and trees take more than 20 years to attain reproductive maturity, and hence controlled crossing from the selected progeny is a tough task. Breeding programs in tree species are hampered due to long lifecycle, polyploidy, complex pollination mechanisms, polygenic controls of desirable characters, self-sterility favouring heterozygous state and natural barriers of interspecific crosses (Burley 1987). Owing to these drawbacks, alternative approaches like tissue culture and genetic engineering techniques can be attempted in order to overcome some of the problems and to achieve faster outcome.

15.7 Tissue Culture

Significant advancement has been made on the optimization of tissue culture techniques with respect to many tree species. Among the various tissue culture tools, the techniques which suffice for the tree improvement are micropropagation from elite genotypes, production of somaclonal variants with desired traits, haploid/triploid/polyploid production, protoplast fusion and gene transfer (Sita and Raghava Swamy 1998). Of the several plant tissue culture techniques, somaclonal variation coupled with in vitro selection to select trees for desirable traits like disease resistance and rapid clonal multiplication of such trees has tremendous potential (Manoj Kumar et al. 1998). However, reports are available only for micropropagation studies, leaving scope for research utilizing other techniques.

Micropropagation is a worth mentioning method which can be adopted for tree improvement as this technology delivers true-to-type plants. This approach will not only provide plantlets with exactly similar to selected plants with desired characteristics like pathogen resistance or better wood quality but will also ensure plantlets availability throughout the year which can be used for large-scale propagation and conservation of the species (Rai and Shekhawat 2014; Kher et al. 2016; Nataraj et al. 2016; Bi et al. 2017; Teixeira da Silva et al. 2017).

Micropropagation studies on red sanders began in the 1980s, mostly confined to usage of seedling explants like shoot tip, node, mesophyll and cotyledons. These explants were subjected to regeneration, and many researchers achieved varying degrees of success (Sarita et al. 1988; Sita et al. 1992; Anuradha and Pullaiah 1999; Arockiasamy et al. 2000; Prakash et al. 2006; Rajeswari and Paliwal 2008; Padmalatha and Prasad 2008; Balaraju et al. 2011; Warakagoda and Subasinghe 2013). In the context of tree improvement of red sanders, though micropropagation offers immense potentiality to generate large number of planting stocks, it is always impressive, provided the starting material is from a selected mature tree explant. However, except a single report (Prakash et al. 2006), all other publications mention the usage of seedling explants. This is the major limitation in tree improvement program via tissue culture. No publications are witnessed about the long-term field trials of tissue-cultured red sanders. Field performance and genetic fidelity studies remain unanswered in *P. santalinus.*

There is a tremendous scope for inducing somaclonal variations, homozygous diploids and polyploid cultures in red sanders. Authors (Indu and Anuradha) in their studies found that mature and immature embryos and reproductive tissue explants are responding well in red sanders. This may be due to very less phenolics and also high regeneration potential. Mutation studies, establishing polyploids through tissue cultures, may be focused, as there are no attempts in these lines. DNA barcoding and marker-assisted selection can be a useful tool to select the plants with desired traits after establishing somaclonal variants and mutants, and this can be the scope for future research in red sanders improvement.

15.8 Molecular and Genetic Engineering

Tree improvement or multiplying selected trees with desired qualities by using seeds or seedlings is not desirable as seeds are heterozygous and may not exhibit anticipated characters. In this regard molecular biology offers various techniques for determination of genetic diversity and assists in selection of genotypes of interest by maker-based technology. Molecular markers are indispensable tools and can be utilized for selection of clones, identification of genetic diversity hotspots, genetic fingerprinting, barcoding, assembly of breeding populations, gene pool monitoring, characterization of population dynamics and gene flow (Arif et al. 2011; Swamy and

Anuradha 2011). There are reports regarding utilization of molecular tools in identification of genetic diversity in red sanders and are discussed as a separate chapter in this monograph. Red sanders tree improvement programs can be benefited from this technology for selection of trees with superior characters without harming the tree. Once standardized the molecular markers specific to elite trees can be employed in identifying at the early stage though the quality of the wood is expressed in the later stages of the tree life cycle. The identification of superior quality *P. santalinus* at their seedling stage by molecular-assisted selection can help the farmer to plant superior trees; otherwise they will be known only after 20 years.

Another promising approach that can be adopted for tree improvement is through genetic engineering wherein desired gene is identified and introduced into the tree genome either through *Agrobacterium*-mediated gene transfer or through biolistic – DNA-coated particle bombardment technique. Genetic modification may be used to alter or introduce wide range of traits including insect resistance, disease resistance, herbicide tolerance and growth rate. Many tree species for timber production have been genetically modified and tested in field trials in several countries, viz. Europe, North and South America and New Zealand. Insect-resistant GM poplar trees have been approved for commercial planting in China. There is enormous potential for speeding up tree breeding cycles by the use of GM technology. However, use of GM technology for tree improvement in red sanders is neither attempted nor practiced. There are absolutely no publications regarding transgenic tree approach in red sanders. However, in the related species as of now, only a single report is available on transient genetic transformation of *P. marsupium* by *Agrobacterium tumefaciens* wherein callus was transformed with *hpt* and *GUS* gene under the control of the CaMV-35S promoter (Tippani et al. 2013). Any successful strategy for tree improvement depends upon the reproducibility, scaling up and field performance. Tree improvement of red sanders is not an exception and more so challenging with many limitations. Every stage of this precious tree is surrounded by hindrances, as reiterated in every chapter of this monograph. Being endemic to restricted areas, the tree is not accessible for studying its phenology and reproduction biology. Slow growth coupled with issues in propagation complicated the scenario. The tree is witnessed to be resilient and recalcitrant for regeneration both for vegetative and micropropagation methods. Vegetative methods are hampered by poor rooting performance and micropropagation by poor response from mature tree explants. Tree improvement through conventional breeding is time-consuming with additional requirement of large area for establishment, storage and testing facilities. Though there are stray attempts about the understanding the genetic makeup and molecular studies, still transgenic technology is not explored. Considering the above facts, serious attempts must be made for further refinement of the strategies in order to revive this valuable species. Tree improvement studies in red sanders must be given impetus and novel modern approaches in some of the areas like seed production, macro and micropropagation techniques, accelerated breeding using marker-assisted selection and functional genomics to be adopted in order to achieve higher productivity and success.

References

Anuradha M, Pullaiah T (1998) Investigations on germination of enhanced axillary branching in *Pterocarpus santalinus* (red sanders) with special reference to in vitro seed culture. Indian For 124:309–314

Anuradha M, Pullaiah T (1999) Propagation studies of red sanders (*Pterocarpus santalinus* L. f.) in vitro- an endangered taxon of Andhra Pradesh, India. Taiwania 44(3):311–324

Arif IA, Khan HA, Bahkali AH, Homaidan AA, Farhan AH, Al Sadoon M, Shobrak M (2011) DNA marker technology for wildlife conservation. Saudi J Biol Sci 18(3):219–225. https://doi.org/10.1016/j.sjbs.2011.03.002

Arockiasamy S, Ignacimuthu S, Melchias G (2000) Influences of growth regulators and explants type on in vitro shoot propagation and rooting of red sandal wood (*Pterocarpus santalinus* L. f.). J Exp Biol 48:1270–1273

Arunkumar AN (2011) Variability studies in *Pterocarpus santalinus* in different aged plantations of Karnataka. My For 47(4):343–353

Balaraju K, Agastian P, Ignacimuthu S (2011) A rapid in vitro propagation of red sanders (*Pterocarpus santalinus* L. f.) using shoot tip explants. Acta Physiologeae Plant 33:2501–2510

Bi WL, Pan C, Hao XY, Cui ZH, Kher MM, Markovic Z, Wang QC, Teixeira da Silva JA (2017) Cryopreservation of grapevine (*Vitis* spp.) – a review. In Vitro Cell Dev Biol Plant 53:449–460

Burley J (1987) Application of biotechnology in forestry and rural development. Commonw For Rev 66:357–367

Comstock RE (1977) Quantitative genetics and the design of breeding programs. In: Pollak E, Kempthorne O, Bailey TB (eds) Proceedings of international conference on quantitative genetics. Iowa State University Press, Ames, pp 705–718

Dayanand T, Lohidas T (1988) Studies on development and maturity of red sanders (*Pterocarpus santalinus* Linn. f.). Indian J For 11:207–208

Jyothi CP, Chandrashekar R, Lakshmi B (2014) Isolation of *Pterocarpus santalinus* L. f. genomic DNA, for quality check and quantification with reference to Telangana region, Andhra Pradesh, India. Indian J Sci 8:21–24

Kalimuthu K, Lakshmanan KK (1995) Effect of different treatments on pod germination of *Pterocarpus* species. Indian J For 18:104

Kedharnath S (1984) Forest tree improvement in India. Proc Indian Acad Sci (Plant Sci) 93 (3):401–412

Kedharnath S, Rawat (1976) Studies on variation in fibre morphology in wavy grained and straight grained trees of red sanders. Indian For 102(7):441–446

Kher MM, Nataraj M, Teixeira da Silva JA (2016) Micropropagation of *Crataeva* L. species. Rend Lincei 27:157–167

Ledig FT (1974) An analysis of methods for the selection of trees from wild stands. For Sci 20:2–16

Lohidas T, Dayanand T (1984) Some distinguishing characteristics between wavy grained and straight grained trees of red sanders. Indian J For 7:69–71

Manoj Kumar M, Reddy M, Nadagoudar B (1998) Application of tissue culture techniques in tree improvement. In: Puri S (ed) Tree improvement – applied research and technology transfer. Oxford and IBH Publishing Co. Pvt. Ltd., New Delhi, pp 249–261

Naidu CV (2000) Seed scarification requirement in *Pterocarpus santalinus* Linn. f. J Indian Bot Soc 79:175–178

Naidu CV (2001a) Improvement of seed germination in red sanders (*Pterocarpus santalinus* Linn. f.) by plant growth regulators. Indian J Plant Physiol 6:205–207

Naidu CV (2001b) Seed pretreatment methods to improve germination in red sanders (*Pterocarpus santalinus*Linn. f.). Indian J For 24:342–343

Naidu CV, Rajendrudu G (2001) Influence of kinetin and nitrogenous salts on seed germination of red sanders (*Pterocarpus santalinus* Linn. f.). Seed Sci Technol 29:669–672

Nataraj M, Kher MM, Teixeira da Silva JA (2016) Micropropagation of *Clerodendrum* L. species: a review. Rend Lincei 27:169–179

Padmalatha K, Prasad MNV (2007) Morphological and molecular diversity in *Pterocarpus santalinus* L.f – an endemic and endangered medicinal plant. Med Aromat Plant Sc Biotechnol 1:263–273

Padmalatha K, Prasad MNV (2008) In vitro plant regeneration of *Pterocarpus santalinus* L.f. (red sanders) – an endangered medicinal plant and important timber tree. Tree For Sci Biotechnol 2 (1):1–6

Prakash E, Shavalikhan PS, Srinivasa Rao TJV, Meru ES (2006) Micropropagation of red sanders (*Pterocarpus santalinus* L.) using nodal explants. J For Res 11(5):329–335

Puri S (1998) Plus-tree selection as a tool in tree improvement. In: Puri S (ed) Tree improvement – applied research and technology transfer. Oxford and IBH Publishing Co. Pvt. Ltd., New Delhi, pp 97–105

Rai MK, Shekhawat NS (2014) Recent advances in genetic engineering for improvement of fruit crops. Plant Cell Tissue Organ Cult 116:1–15

Rajeswari V, Paliwal K (2008) In vitro plant regeneration of red sanders (*Pterocarpus santalinus* L. f.) from cotyledonary nodes. Indian J Biotechnol 7:541–546

Rani JS, Usha R (2013) Development of RAPD and specific SCAR markers for the identification of *Pterocarpus santalinus* L. J Cell Tissue Res 13(3):3809–3816

Rao SP, Raju AJS (2002) Pollination ecology of the red sanders *Pterocarpus santalinus* (Fabaceae), an endemic and endangered tree species. Curr Sci 83:1144–1148

Rawat MS, Uniyal DP (1996) Identification of wavy grained red sanders (*Pterocarpus santalinus*) at nursery stage. Indian For 122:9–12

Reddy KK, Srivasuki KP (1990) Vegetative propagation of red sanders (*Pterocarpus santalinus* Linn. f.). Indian For 116(7):536–540

Reddy KK, Srivasuki KP (1992) Biotechnological approach for tree improvement in red sanders. Indian For 118:15–20

Sarita S, Bhatnagar SP, Bhojwani SS (1988) Preliminary investigations on micropropagation of leguminous timber tree *Pterocarpus santalinus*. Phytomorphology 38:41–45

Sita GL, Swamy R (1998) Application of biotechnology in forest trees – clonal multiplication of sandalwood, rosewood, eucalyptus, teak and bamboos by tissue culture in India. In: Puri S (ed) Tree improvement – applied research and technology transfer. Oxford and IBH Publishing Co. Pvt. Ltd., New Delhi, pp 233–246

Sita GL, Sreenatha KS, Sujata S (1992) Plantlet production from shoot tip cultures of red sandal wood (*Pterocarpus santalinus* L. f.). Curr Sci 62(7):532–535

Soundararajan V, Joshi SC (2012) Tree improvement. In: Endemic possessions of Eastern Ghats: red sanders (*Pterocarpus santalinus* Linn. f). Institute of Wood Science and Technology, (ICFRE), Bangalore, pp 23–38

Swamy MK, Anuradha M (2011) Analysis of genetic variability in Patchouli cultivars (*Pogostemoncablin* Benth.) by using RAPD markers. Res Biotechnol 2(6):64–71

Teixeira da Silva JA, Zeng S, Wicaksono A, Kher MM, Kim H, Hosokawa M, Dewir YH (2017) In vitro propagation of African violet: a review. South Afr J Bot 112:501–507

Teixeira da Silva JA, Kher MM, Soner D, Nataraj M (2018) Indian Kino tree (*Pterocarpus marsupium*): propagation, micropropagation and biotechnology. Environ Exp Biol 16:1–8

Tippani R, Yarra R, Bulle M, Porika M, Abbagani S, Thammidala C (2013) In vitro plantlet regeneration and *Agrobacterium tumefaciens*-mediated genetic transformation of Indian Kino tree (*Pterocarpus marsupium* Roxb.). Acta Physiologea Plant 35:3437–3446

Warakagoda P, Subasinghe S (2013) In vitro propagation of *Pterocarpus santalinus* L. f. (red sandalwood) through tissue culture. J Natl Sci Found Sri Lanka 41(1):53–63

The manufacturer's authorised representative in the EU is Springer Nature Customer Service Centre GmbH, Europaplatz 3, 69115 Heidelberg, Germany. If you have any concerns regarding our products, please contact ProductSafety@springernature.com

Printed and bound by CPI Group (UK) Ltd, Croydon, CR0 4YY
30/09/2026
02225235-0006